Essential Questions to Master the California Exam

Two Practice Exams

First Edition

AF758678

JACOB PETRO
PhD, PMP, CEng, PE

Essential Questions to Master the California Seismic Principles Exam

Two Practice Exams

First Edition

JACOB PETRO
PhD, PMP, CEng, PE

PETRO PUBLICATIONS
HILLSBORO BEACH

For the most up-to-date list of errata, or for any other information please contact us at

PEguides@PetroPublications.com

PETRO
PUBLICATIONS

Report Errors For this Book

We are grateful to every reader who notifies us of possible errors. Your feedback allows us to improve the quality and accuracy of our products.

Report errata by sending an email to PEguides@PetroPublications.com

Essential Questions to Master the California Seismic Principles Exam
Two Practice Exams

First Edition Print 1.02

© 2025 Petro Publications LLC. All rights reserved.

All content is copyrighted by Petro Publications LLC and its owners. No part, either text or image, may be used for any purpose other than personal use. Reproduction, modification, storage in retrieval system or retransmission, in any form or by any means, electronic, mechanical, or otherwise, for reasons other than personal use, without prior permission from the publisher is strictly prohibited.

For general inquiries contact: Admin@PetroPublications.com

Imprint name: PE Essential Guides

Company owning this imprint: Petro Publications LLC. Established in Florida, 2023.

ISBN: 979-8-9921358-1-7

Disclaimer

The information provided in this book is intended solely for educational and illustrative purposes. It is important to note that the technical information, examples, and illustrations presented in this book should not be directly copied or replicated in real engineering reports or any official documentation.

While there may be resemblances between the examples in this book and real structures, users must exercise caution and conduct comprehensive verification of all information before implementing it in any practical setting. The author and all affiliated parties explicitly disclaim any responsibility or liability arising from the misuse, misinterpretation, or misapplication of the information contained in this book.

Furthermore, it is essential to understand that this book does not constitute legal advice, nor can it be considered as evidence or exhibit in any court of law. It is not intended to replace professional judgment, and readers are encouraged to consult qualified experts or seek legal counsel for any specific legal or technical matters.

By accessing and utilizing the information in this book, readers acknowledge that they do so at their own risk and agree to hold the author and all affiliated parties harmless from any claims, damages, or losses resulting from the use or reliance upon the information provided herein.

Important Information About Printing and Quality Assurance

Most of our publications are printed through third-party services. While we have full trust in our partners, we acknowledge that occasional minor issues, such as missing pages or other errors, may arise. If you encounter any such issues with your copy, please do not hesitate to contact us via any of the emails listed on the copyright page, and we will be happy to assist you in resolving the matter.

This page is intentionally left blank

About the Author

Dr. Petro is a professional engineer and a business leader with over 20 years of experience in leading and growing engineering companies. Throughout his career, he worked with some of the most prestigious engineering firms. With a vast background in design and construction, he earned a reputation for delivering innovative and cutting-edge projects throughout his career. Dr. Petro is the author of the PE Essential Guides series, a best-selling collection that provides extensive coverage across all civil engineering disciplines.

Dr. Petro is a civil engineer, he holds a Doctorate degree, he has earned a Professional Engineering (PE) license as well as Chartered Engineering (CEng) certification. Additionally, Dr. Petro has earned a Project Management Professional (PMP) certification, further demonstrating his expertise in managing complex projects. Over the years, he successfully led and managed teams of engineers, designers, and other professionals, overseeing complex projects from conception to completion.

Throughout his career, Dr. Petro designed and delivered numerous innovative and interesting projects that have contributed significantly to various industries he has worked in. His passion for engineering and business has driven him to publish several papers and articles in industry-leading journals and magazines. His work has been recognized as state-of-the-art and has been referenced by many industry professionals.

As an international civil engineer who worked across the globe, Dr. Petro brings an interesting perspective to the table. He has a deep understanding of how civil facilities and structures work and how to optimize them for maximum efficiency and safety. His ability to communicate complex engineering concepts to both technical and non-technical stakeholders has been key to his success.

This page is intentionally left blank

Important Overview of this Book and *The Essential Guide*

About This Book

This book contains two full-length practice exams specifically designed to help you master the California Seismic Principles Exam. The questions closely reflect the actual exam in both length and difficulty and are carefully crafted based on the latest exam specifications. Each solution includes detailed explanations and direct references to the official exam materials, ensuring that you gain a deep understanding of the core concepts.

This book is also designed to complement our main reference guide: *The Essential Guide to Passing the California Seismic Principles Exam – A Reference Manual.* While the practice exams are fully compatible with our reference guide, it is not required to obtain it to use this book effectively. If you have obtained this book as a standalone resource for extra practice, and you do not wish to own our reference manual *"The Essential Guide"*, you will still benefit from this book alone.

If you did not own our publication *The Essential Guide*, and you were using this book as a stand-alone practice book, then you will find that each solution in this book includes citations from the official exam reference list, clearly presented in grey text boxes at the end of each answer. These citations guide you to the exact sources used.

These practice questions were created with your success in mind. They provide comprehensive coverage of the exam topics and are structured to reinforce key concepts. When combined with the 160 practice questions included in our reference manual *The Essential Guide*, you will have access to a total of 270 high-quality problems – this provides you with a complete set that ensures you are fully prepared for any topic that may appear on the exam.

For your peace of mind, both this book and *The Essential Guide* have been tested in real exam scenarios, with full respect to exam copyrights.

The Essential Guide

Throughout the solutions in this book, you will notice frequent references to our main publication, *The Essential Guide to Passing the California Seismic Principles Exam – A Reference Manual*, which we refer to simply here as *The Essential Guide*. While it is not required to own *The Essential Guide* to complete the practice questions in this book, we highly recommend it if you find yourself needing further clarification or more comprehensive coverage of certain topics that may appear on the exam.

The Essential Guide is a complete, exam-ready reference manual that you can take with you on test day. It has been tested by numerous examinees, both before and during the actual exam, and has proven to be a valuable tool for success.

The Essential Guide is organized into seven chapters that align closely with the most recent exam specifications. Inside it, you will find in-depth explanations of key concepts in seismology and code applications that go beyond what is typically found as scattered information inside codebooks or manuals. It clarifies not just the what, but also the why and how behind important topics. This includes guidance on using resources like SDPWS 2021, CBC 2022, ASCE 7-16, ATC-20, TMS 402/602-16 and others, along with the design and required detailing and specifications for various construction material, such as wood, masonry, steel and reinforced concrete.

Additionally, *The Essential Guide* features proprietary processes and design diagrams that simplify complex topics – such as horizontal and vertical irregularities, design of nonstructural components and nonbuilding structures, and the treatment of rigid versus flexible diaphragms, and many others. Rather than having to piece together this information from various sources, *The Essential Guide* presents it in a clear, streamlined, format tailored for quick reference.

In *The Essential Guide*, you will also find summary tables covering seismic requirements from the above-mentioned codes and standards. These tables are designed for fast access during the exam, reducing the time spent flipping through multiple documents. Furthermore, *The Essential Guide* includes key tables from the ASCE 7 manual, reprinted with official permission from the relevant code owning authorities, so you can have the most important references at your fingertips – all in one place.

Important Disclaimer:
None of our publications copies nor replicates any proprietary content from the actual exam, or any other examination material. All material presented here is original, developed to provide clear, step-by-step explanations that help you build true understanding. These resources are designed to equip you with the knowledge and confidence needed to pass the exam through your own preparation and mastery.

Required References

The California Seismic Principles Exam requires you to thoroughly study six references as noted below. It is also recommended to check the BPELSG website from time to time to verify updates to this list.

- 2022 California Building Code (CBC) California Code of Regulations, Title 24, Part 2, Volumes 1 and 2.
  You can access it online using: https://codes.iccsafe.org
- American Society of Civil Engineers, ASCE/SEI 7-16 Minimum Design Loads for Buildings and Other Structures.
- AWC SDPWS 2021 Special Design Provisions for Wind and Seismic.
  You can access it online using: https://awc.org/publications/2021-sdpws
- TMS 402/602-16 Building Code Requirements and Specifications for Masonry Structures.
- ATC 20 Procedures for Postearthquake Safety Evaluation of Buildings and ATC 20-1 Field Manual, 2nd Edition, 1995.
- Design of Wood Structures ASD/LRFD 7th Edition, 2014 by Donald E. Breyer, Kenneth J. Fridley and Kelly E. Cobeen. McGraw-Hill, Inc. New.

While *The Essential Guide to Passing the California Seismic Principles Exam* summarizes and includes key tables (with the necessary permissions) from some of the abovementioned key references, these original references contain hundreds of pages, and the guide is not intended to summarize every detail. Additionally, some of these references are available online, where you can obtain permission for a nominal fee to print and bring a copy with you to the exam for your own use only.

The Exam

At the time of publishing this book, the exam consists of 55 questions, with candidates given 150 minutes to complete them. This averages to just under three minutes per question. It is important to remember that while some questions may take only a minute to solve, others could take up to five or six minutes. To prepare effectively, practice as many problems as possible beforehand so you can quickly locate relevant information during the exam without wasting time.

IMPORTANT: Even if you are well-versed in the exam material and fully understand the concepts, a lack of familiarity with the exam structure – or not knowing where to quickly locate key equations, tables, or figures – can significantly impact your performance. Time management is critical, and without this familiarity, you may struggle to complete the exam within the allotted time. This is a situation you should avoid.

Exam Structure

The structure of the seismic principles exam aligns with the Board for Professional Engineers, Land Surveyors, and Geologists (BPELSG) test plan for 2022.

Below, you will find the test plan, including the weight and the possible number of questions per section.

	Title	**Exam Weight**	**Possible No. of Questions**
I	Seismic Data and Seismic Design Criteria	8%	4 *to* 5
II	Seismic Characteristics of Engineered System	8%	4 *to* 5
III	Seismic Vulnerability and Improvement of Structural Systems	6%	3 *to* 4
IV	Seismic Forces: Building Structures	26%	14 *to* 15
V	Seismic Forces: Non-Building Structures, Components & Equipment	14%	7 *to* 8
VI	Seismic Analysis Procedure	26%	14 *to* 15
VII	Seismic Detailing and Construction Quality Control	12%	6 *to* 7

Tips for Success

After (or during) completing each practice test, make it a habit to record any areas of weakness in the designated *Notes* section next to each question on the answer sheet, or in the *Notes* section provided at the end of the book. Additionally, if you come across important concepts or insights that you want to remember or revisit during your exam preparation, jot them down in the *Notes* section as well. This will serve as a valuable reference during your final review and help reinforce key information.

This page is intentionally left blank

EXAM ONE

Answer Sheet

No.	Your Answer	Fill our Answer	Notes
1.1			
1.2			
1.3			
1.4			
1.5			
1.6			
1.7			
1.8			
1.9			
1.10			
1.11			
1.12			
1.13			
1.14			
1.15			
1.16			
1.17			
1.18			
1.19			
1.20			
1.21			
1.22			
1.23			
1.24			
1.25			
1.26			
1.27			
1.28			
1.29			
1.30			
1.31			
1.32			
1.33			
1.34			
1.35			
1.36			
1.37			
1.38			
1.39			
1.40			
1.41			
1.42			
1.43			
1.44			
1.45			
1.46			
1.47			
1.48			
1.49			
1.50			
1.51			
1.52			
1.53			
1.54			
1.55			

PROBLEM 1.1
A site is being evaluated for seismic design and the following information was collected based on field tests:

- First $35\ ft$ average undrained shear strength = $1{,}400\ psf$
- Second $65\ ft$ average undrained shear strength = $800\ psf$
- Third $50\ ft$ average undrained shear strength = $1{,}750\ psf$

Based on the above information, what is the site classification for this location?

(A) Class C – Dense soil or soft rock

(B) Class D – Stiff soil

(C) Class E – Soft clay soil

(D) Class F – Vulnerable to failure

PROBLEM 1.2
Which of the following soil type has low potential for liquefaction in coastal areas?

(A) Marine clay.

(B) Coastal reclaimed land and artificial fill.

(C) Soft unconsolidated marine deposits.

(D) Loose, saturated sands with silts in coastal areas.

PROBLEM 1.3
A site is being evaluated for seismic design based on the cross-section provided below, which illustrates one of the individual footings for the structure to be constructed.

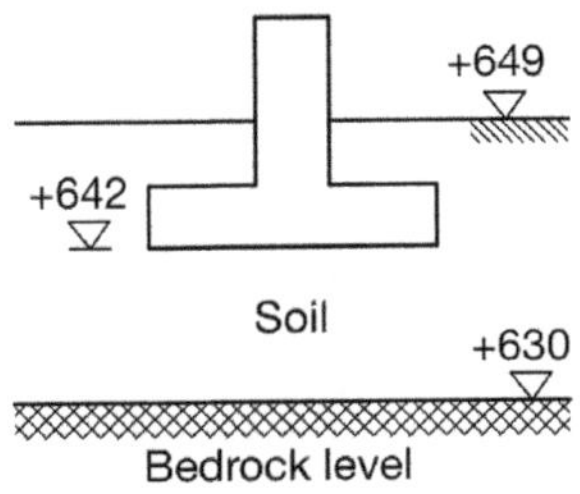

The soil layer shown consists of compacted soil with shear wave velocity of $1{,}800\ ft/s$, while the bedrock has a shear wave velocity of $5{,}000\ ft/s$.

Based on the above information, what is the site classification for this location?

(A) Class A

(B) Class B

(C) Class C

(D) Class D

PROBLEM 1.4
A police station that will be located in a certain site that did not undergo any site investigations yet, along with the following data which were collected from the ASCE online hazard tool:

- $S_S = 0.25$
- $S_1 = 0.10$

Based on the above information, what is the best Seismic Design Category (SDC) for this building?

(A) SDC C

(B) SDC D

(C) SDC E

(D) SDC F

PROBLEM 1.5
A new retail building is planned to be $65\ ft$ tall and will include multiple public assembly areas, with a total expected occupant load of 2,000. The following information has been gathered for the proposed structure location:

- $S_S = 0.75$
- $S_1 = 0.25$
- Site Class C

Based on the above information, what is the best Seismic Force Resisting System (SFRS) for Bearing Wall System that can be provided for this building?

(A) Ordinary reinforced concrete shear walls.

(B) Light wood frame walls sheathed with wood structural panels rated for shear resistance.

(C) Intermediate precast shear walls.

(D) Ordinary plain masonry shear walls.

PROBLEM 1.6

Which one of the below systems is best fit to build a 65 ft tall residential building in an SDC E with the following data?

- $S_S = 0.75$
- $S_1 = 0.25$

(A) Special reinforced concrete shear walls.

(B) Ordinary reinforced concrete shear walls.

(C) Light frame (cold-formed steel) walls sheathed with wood structural panels rated for shear resistance.

(D) Intermediate precast shear walls.

PROBLEM 1.7

Approximately, what is the natural frequency for a 12 story building, with each story 14 ft tall, constructed using steel moment-resisting frames?

(A) 0.9 Hz

(B) 2 sec

(C) 0.6 Hz

(D) 1 sec

PROBLEM 1.8

For a single-degree-of-freedom structure with steel bracing, what is the effect on the natural frequency (f) if additional braces are added, doubling the structure's stiffness while keeping its mass unchanged?

(A) $f_{new} = \sqrt{2} \times f_{current}$

(B) $f_{new} = 2 \times f_{current}$

(C) $f_{new} = \frac{1}{\sqrt{2}} \times f_{current}$

(D) $f_{new} = \frac{1}{2} \times f_{current}$

PROBLEM 1.9

To enhance the ductility of an ordinary reinforced masonry shear wall in a seismic force-resisting system, which of the following strategies could be considered?

(A) Increase area and improve spacing of reinforcements.

(B) Reduce number of openings in the wall.

(C) Increase number of openings in the wall.

(D) Use higher steel grade in the wall reinforcement.

PROBLEM 1.10

What should be the minimum separation distance between the following two buildings in SDC B?

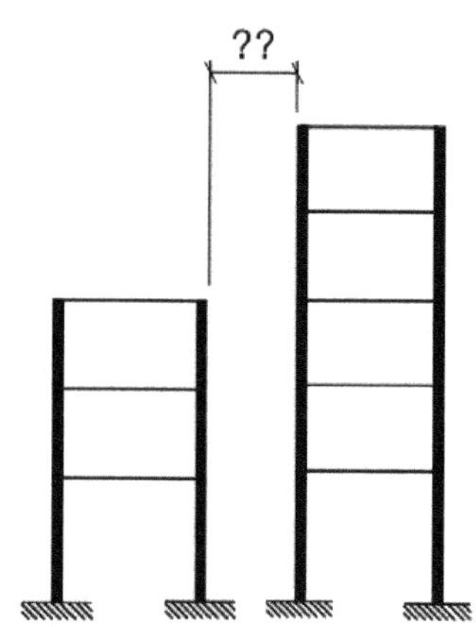

Building 1 (left):

- Building with public assemblies with occupant load of 250 per assembly, and a total occupant load of 2,000.

- The building is made of steel and concrete composite ordinary moment frame.

Building 2 (right):

- A mall with an occupant load of 5,500.
- The building is made of intermediate reinforced concrete moment frame.

Elastic deflections for the two buildings are as follows:

Floor No	Building 1	Building 2
1	1.9 in	1.7 in
2	2.8 in	2.7 in
3	3.5 in	3.7 in
4	–	4.6 in
5	–	5.5 in

(A) 7.2 in

(B) 12.3 in

(C) 15.9 in

(D) 38.0 in

PROBLEM 1.11

The following sketch is for a flexible diaphragm showing elastic deflections as measured using a computer model.

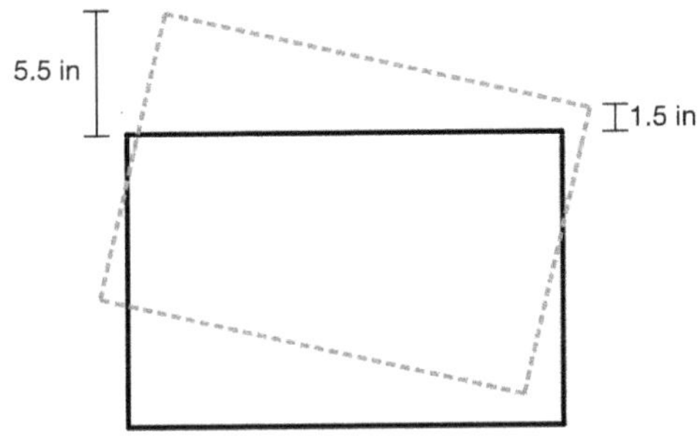

If this structure was assigned to a Seismic Design Category C, what would be the amplification factor for the accidental torsional moment that should be used?

(A) 0

(B) 1.1

(C) 1.7

(D) 3.0

PROBLEM 1.12

What is the maximum height for a weak story that can be accepted without the need to provide any modifications for seismic forces?

(A) 25 ft

(B) 30 ft

(C) 35 ft

(D) 40 ft

PROBLEM 1.13

According to ASCE 7, the vertical distribution of seismic base shear in a structure is primarily based on which mode of vibration?

(A) Third mode

(B) Second mode

(C) First mode

(D) A combination of several modes

PROBLEM 1.14

Which type of building is most commonly observed to have two parts of the building separate during a seismic event?

(A) Precast concrete buildings

(B) Tilt-up wall structures

(C) Reinforced masonry buildings

(D) Small wood-frames buildings

PROBLEM 1.15

To prevent buckling and ensure the proper transfer of wall anchorage forces across a metal deck diaphragm, especially in the direction perpendicular to the deck corrugations, what is the recommended spacing interval for supports such as joists connected to anchorage-resisting walls?

(A) 4 ft to 6 ft

(B) 6 ft to 8 ft

(C) 8 ft to 10 ft

(D) 10 ft to 12 ft

PROBLEM 1.16
A tilt-up concrete wall is anchored to a flexible diaphragm and includes a $4\,ft$ parapet extension above the roof level, as shown.

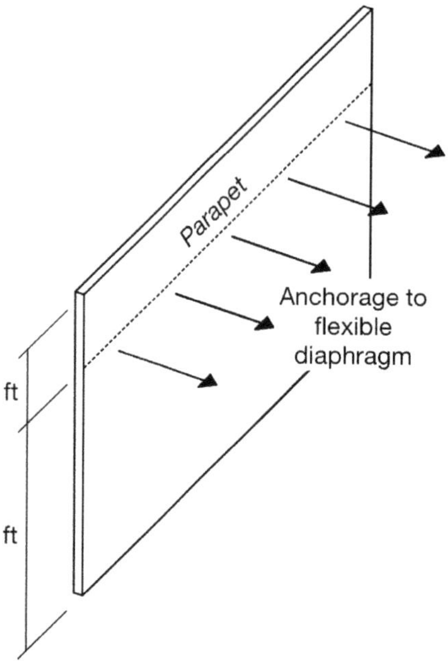

The wall is part of a fire station. The wall thickness is $9\,in$, and the unit weight of concrete is $150\,pcf$. The Site Class is C, with seismic parameters $S_s = 0.9$ and $S_1 = 0.3$.

Given that the wall extends $80\,ft$ between vertical rigid supports that provide lateral support to the flexile diaphragm, what is the linear seismic anchorage force at the roof anchor level (in lb/ft or plf)?

(A) $375\,plf$
(B) $475\,plf$
(C) $675\,plf$
(D) $875\,plf$

PROBLEM 1.17
A warehouse used for furniture storage, located on the second floor of a retail facility, has an effective dead weight of $175\,kip$ (excluding stored items). The floor area measures $45\,ft$ by $25\,ft$, and it is expected to store no more than $7.5\,kip$ of furniture. The live load for this floor is designed as $125\,psf$, in accordance with ASCE 7 Table 4.3-1.

Based on this information, what is the total live load that should be included in the calculation of the final effective seismic weight?

(A) $0\,kip$
(B) $35\,kip$
(C) $70\,kip$
(D) $140\,kip$

PROBLEM 1.18
What is the design story drift for a $12\,ft$ tall story for eleven stories building constructed entirely using moment-resisting frames, located in SDC D, classified as a critical national defense facility, and exhibiting a Type 1b horizontal structural irregularity?

(A) $1.11\,in$
(B) $1.44\,in$
(C) $1.66\,in$
(D) $2.16\,in$

PROBLEM 1.19
Which from the following cross-sections best illustrates the shear transfer mechanism between a wood diaphragm and a masonry shear wall?

(A) Option A

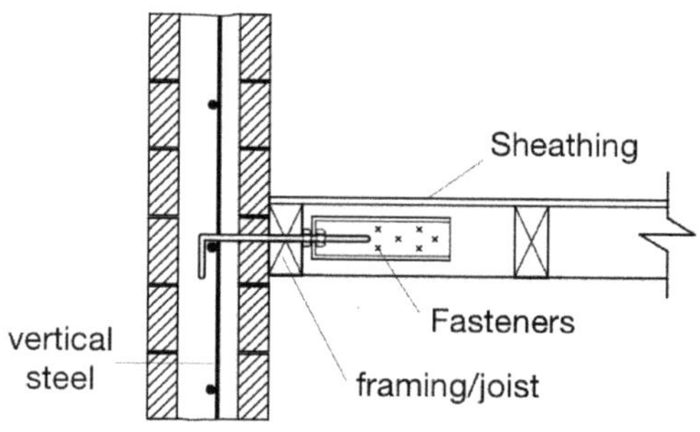

(B) Option B

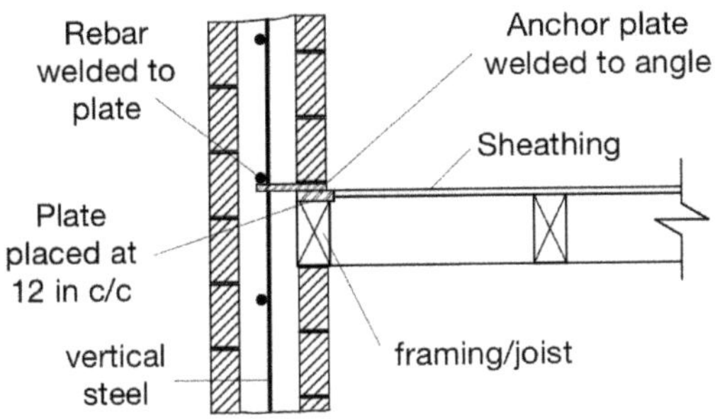

(C) Option C

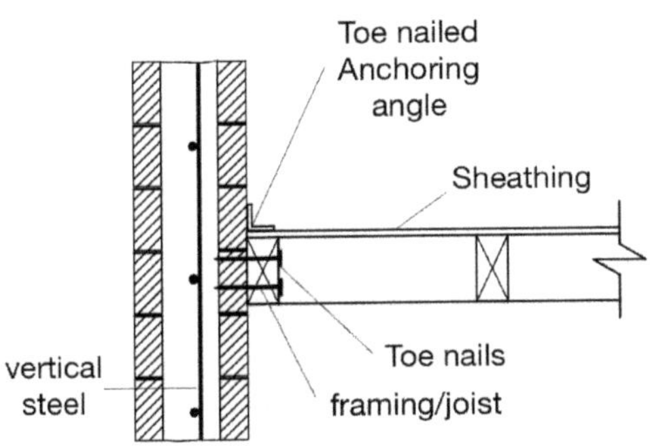

(D) Option D

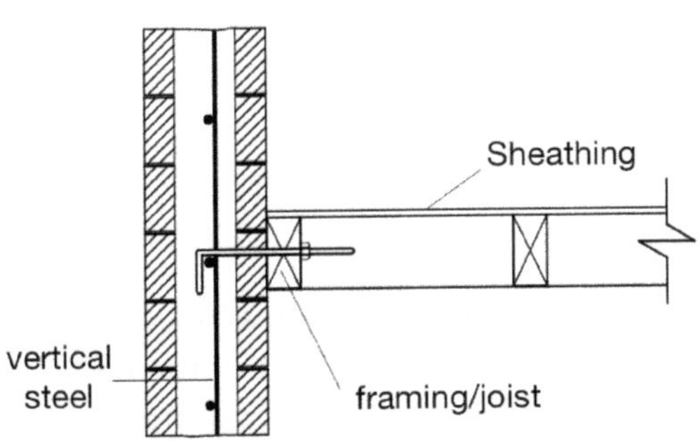

PROBLEM 1.20

To enhance the ductility of the seismic force-resisting system (SFRS) in a specific direction, ordinary plain AAC masonry shear walls – configured as a bearing wall system – have been supplemented with steel eccentrically braced frames in that direction.

Given this combination, what would be the new response modification factor for the integrated system in that direction?

(A) 9.5

(B) 8

(C) 4.75

(D) 1.5

PROBLEM 1.21

For the rigid diaphragm shown below, assume the origin of the x- and y-axes is located at the bottom-left corner. What are the x and y coordinates of the center of rigidity (*)?

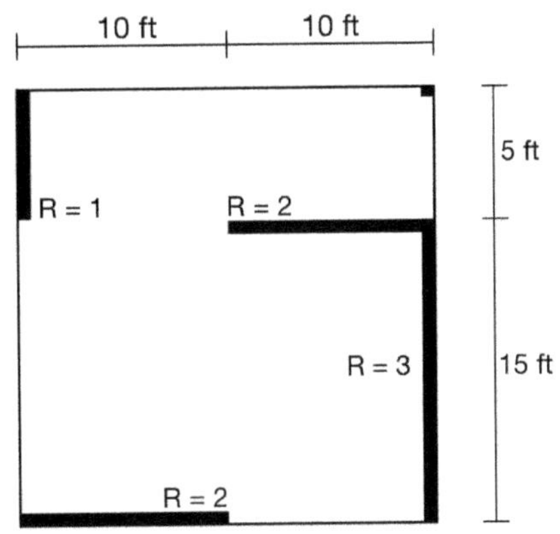

(A) (7.5, 13)

(B) (15, 7.5)

(C) (13, 10)

(D) (13, 7.5)

(*) Assume that the orthogonally connected shear walls are not structurally connected, as considering their combined stiffness is beyond the scope of this book.

PROBLEM 1.22

For the rigid diaphragm shown below, assume the origin of the x- and y-axes is located at the bottom-left corner. All walls are 12 *in* thick and 16 ft tall concrete walls, all walls can be assumed as cantilever walls. What are the x and y coordinates of the center of rigidity (*)?

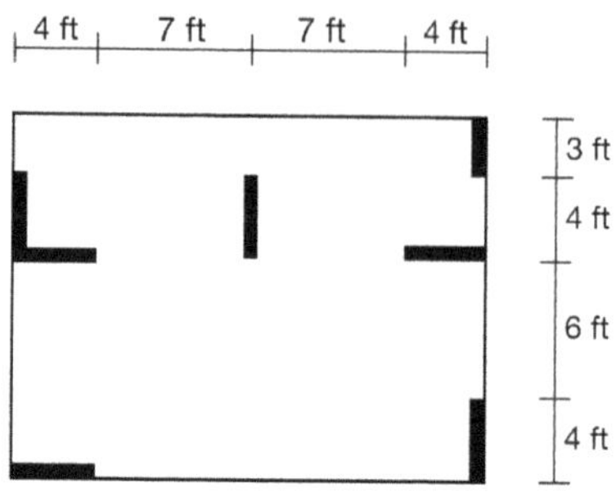

(A) $(13.5, 7.5)$
(B) $(12.4, 6.7)$
(C) $(11, 5.7)$
(D) $(11, 6.7)$

(*) Assume that the orthogonally connected shear walls are not structurally connected, as considering their combined stiffness is beyond the scope of this book.

PROBLEM 1.23

The building shown below, is made of steel and concrete composite special shear walls as an SFRS, the effective seismic weight is indicated at each diaphragm, and each story heigh is $12\,ft$.

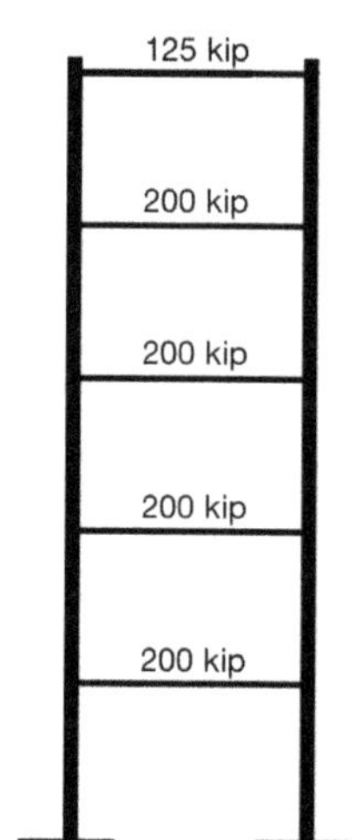

Using the following seismic coefficients: $S_{DS} = 0.6$, $S_{D1} = 0.8$, $T = 0.45$, and an importance factor of 1.0, what is the required diaphragm design force at the third level?

(A) $21\,kip$
(B) $24\,kip$
(C) $27\,kip$
(D) $34\,kip$

PROBLEM 1.24

A 12-story building uses steel and concrete composite special concentrically braced frames as its SFRS. Each story is $10\,ft$ tall. Use the following seismic parameters for this building:

- $S_S = 0.35$
- $S_1 = 0.65$
- $F_a = 0.9$
- $F_v = 0.8$
- $I_e = 1.0$

Using this information, what is the seismic response coefficient C_s that should be used for this building to calculate its base shear?

(A) 0.009
(B) 0.042
(C) 0.065
(D) 0.095

PROBLEM 1.25

A three-story steel building with no irregularities is located in SDC B. The building includes balconies on both the second and third floors, each $30\,ft$ wide.

What is the maximum allowable cantilever projection of these balconies for them to qualify for the simplified seismic procedure?

(A) $6\,ft$
(B) $10\,ft$
(C) $12\,ft$
(D) $18\,ft$

PROBLEM 1.26

A steel storage rack with two levels is modelled using the forces presented in the following table.

Level	Force (kip)	Level Displacement (in)	Seismic Weight (kip)
1	5	1.7	73
2	17	2.2	73

What is the fundamental period for this nonbuilding structure?

(A) 1.12 sec

(B) 0.88 sec

(C) 0.47 sec

(D) 0.18 sec

PROBLEM 1.27

The six sketches below illustrate different configurations of a second-floor wooden diaphragm connection to an external wooden shear wall, with floor joists running parallel to the wall. Select the three configurations that provide an effective and continuous lateral and vertical load path.

(A) Option A

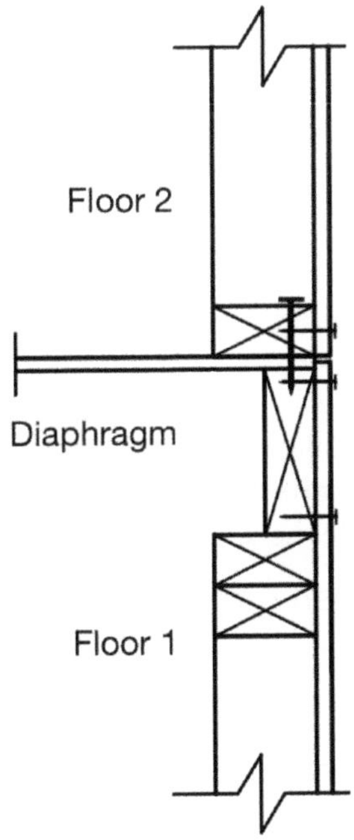

(B) Option B

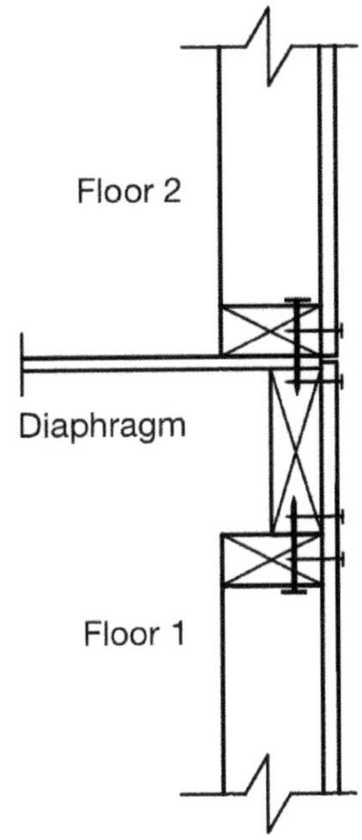

(C) Option C

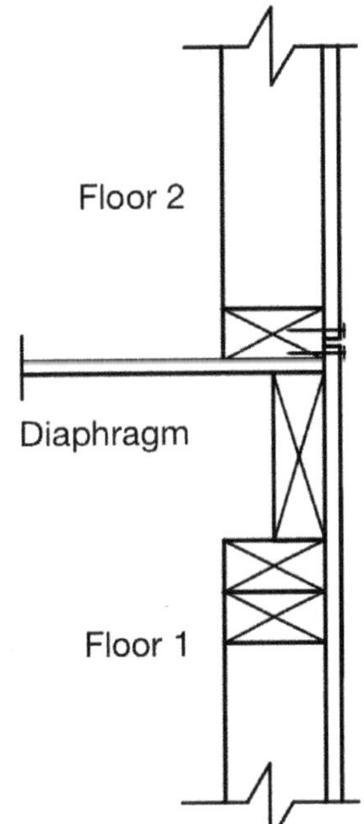

(D) Option D

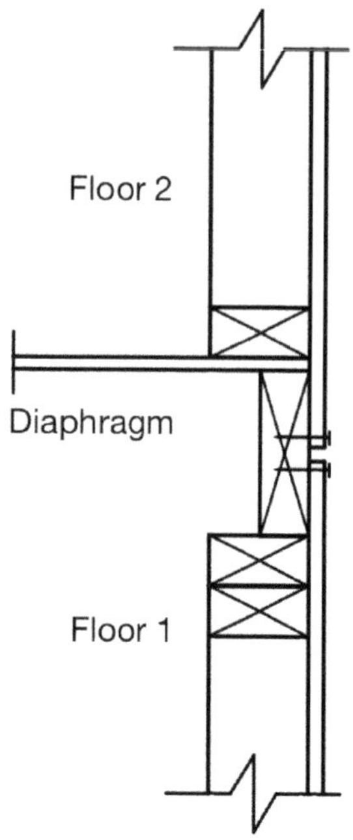

(E) Option E

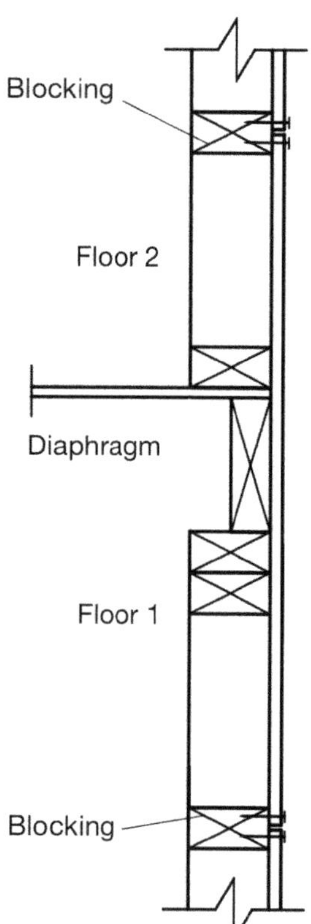

(F) Option F

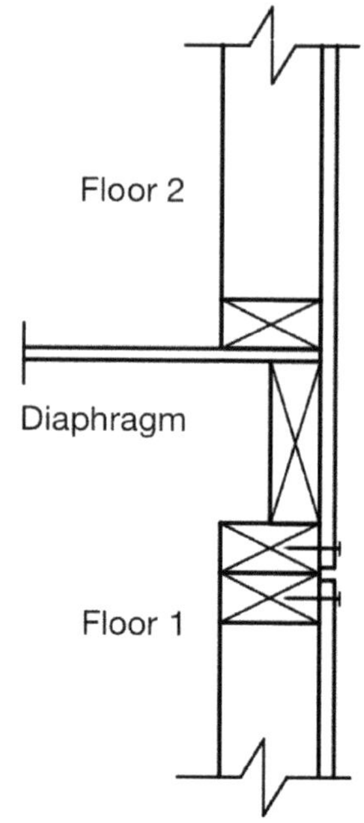

PROBLEM 1.28

A structural element that is part of an ordinary reinforced concrete moment frame supports a building with an out-of-plane offset irregularity, that is, the Seismic Force Resisting System of the upper floor does not align with that of the lower floor.

The structural element is subject to the following loads:

- Dead load (D)
- Live load (L)
- Seismic horizontal load effect (Q_E)

Given:

- $S_{DS} = 0.5$
- $S_{D1} = 0.8$

Based on this information, which equation best represents the LRFD load combination that includes seismic effects?

(A) $1.15D + 3Q_E + L$

(B) $0.8D + 3Q_E$

(C) $0.8D + 2.5Q_E$

(D) $0.95D + 3Q_E + L$

PROBLEM 1.29

Choose only three of the following conditions where overstrength must be used when considering horizontal load effects in load combinations with horizontal earthquake effects:

(A) Foundation and other elements providing overturning resistance at the base of cantilever column systems.

(B) Structural elements supporting walls in buildings with horizontal irregularity Type 2 or vertical irregularity Type 3.

(C) Collector elements in Seismic Design Categories C through F.

(D) Lateral bracing for architectural components not part of the primary structural frame.

(E) Batter piles and their connections.

(F) Elevator shaft walls designed for out-of-plane forces.

PROBLEM 1.30

What is the tension in the chord at the point indicated on the sketch below?

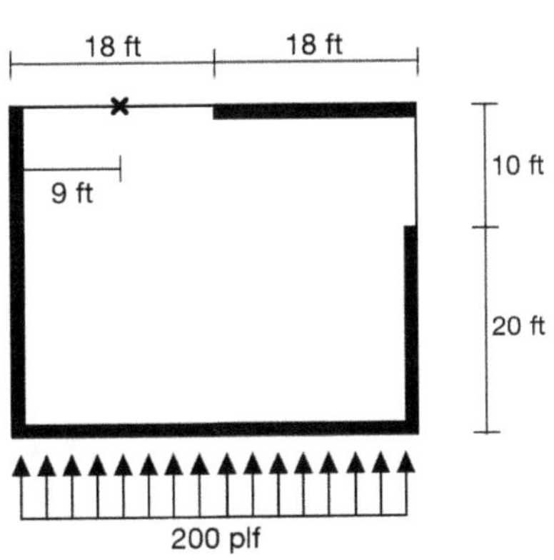

(A) 0.8 *kip*

(B) 1.2 *kip*

(C) 1.5 *kip*

(D) 2.2 *kip*

PROBLEM 1.31

The sketch below shows a wood roofing diaphragm connected to a wood shear wall, with the main diaphragm framing perpendicular to the wall.

In this arrangement, which member acts as the chord of the diaphragm?

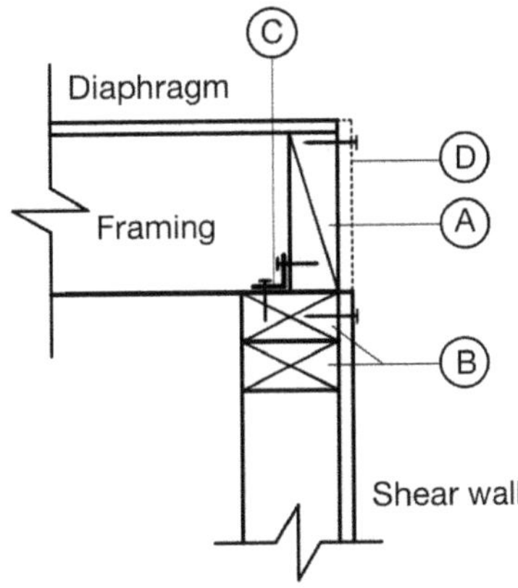

(A) The blocking of the main frame.

(B) The double top plates of the wall.

(C) A fastened long steel L section.

(D) An extension of the wall sheathing to cover the blocking.

PROBLEM 1.32

The below is a reinforced concrete diaphragm along with an edge beam. Identify the best location for chord tension reinforcement.

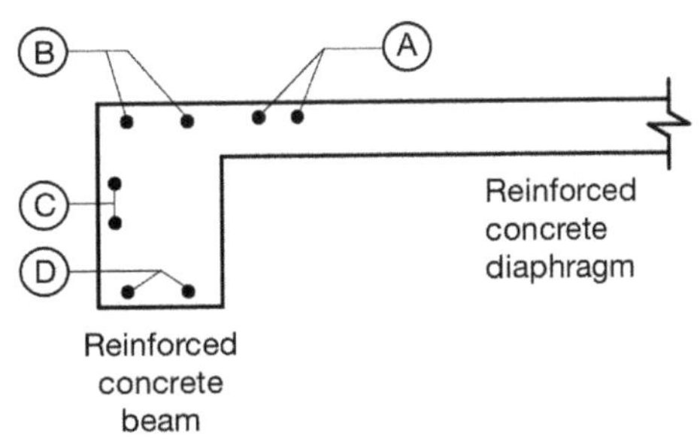

(A) Option A

(B) Option B

(C) Option C

(D) Option D

PROBLEM 1.33

The below two concrete shear walls are fixed at the base and support a one-story flexible diaphragm – i.e., free at the top.

What is the lateral load share for the wall on the left?

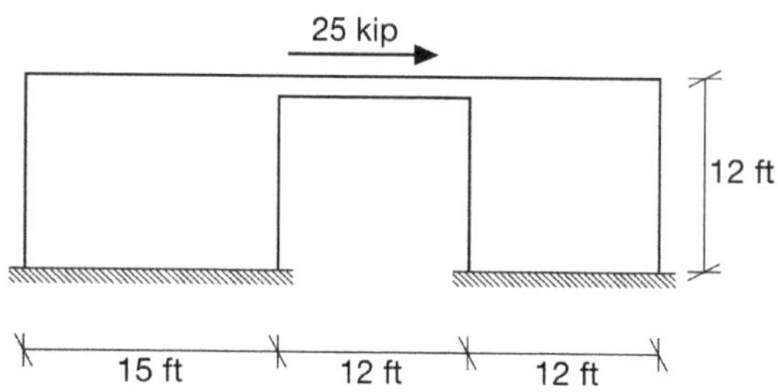

(A) 12.5 *kip*

(B) 13.8 *kip*

(C) 15.3 *kip*

(D) 17.1 *kip*

PROBLEM 1.34

For the below single story, flexible diaphragm supported by shear walls, what is the maximum drag force experienced by the collector circled on the below sketch?

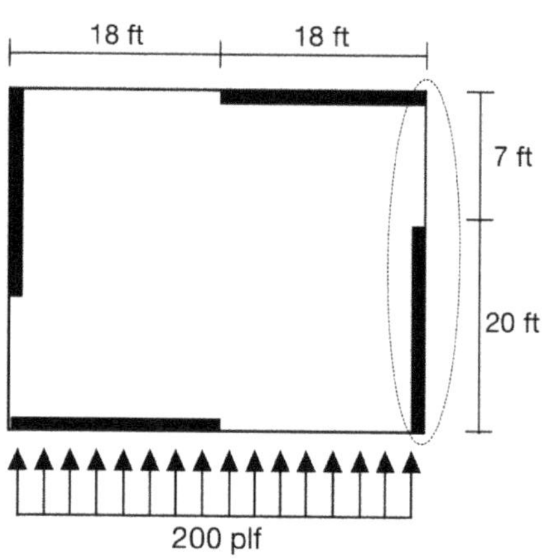

(A) 0.93 *kip*

(B) 1.86 *kip*

(C) 3.63 *kip*

(D) 4.11 *kip*

PROBLEM 1.35

For the below single story, flexible diaphragm supported by wood shear walls, what is the maximum drag force experienced by the collector circled on the below sketch?

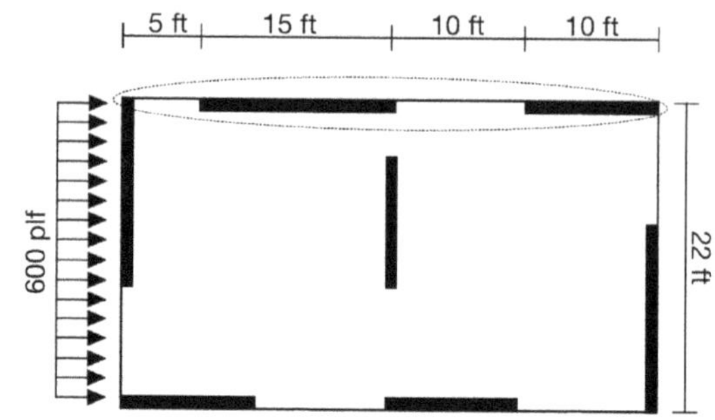

(A) 0.52 *kip*

(B) 0.66 *kip*

(C) 0.83 *kip*

(D) 0.99 *kip*

PROBLEM 1.36

For the below single story, flexible diaphragm supported by shear walls, what is the load share for the wall at axis C for a diaphragm force of 50 *kip* acting in the north direction?

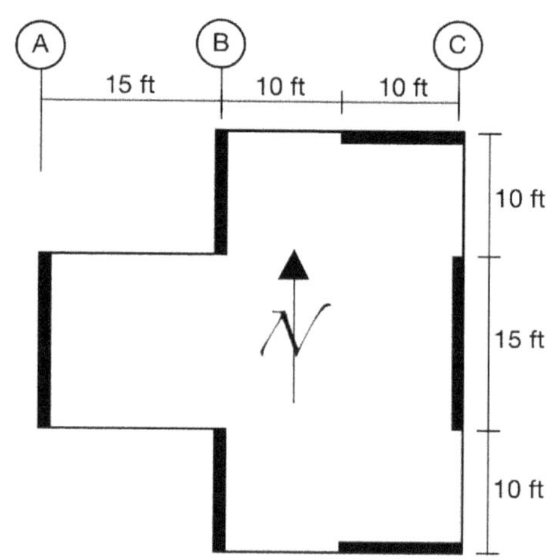

(A) 14.2 *kip*

(B) 18.9 *kip*

(C) 22.9 *kip*

(D) 29.6 *kip*

PROBLEM 1.37

For the below single story, flexible diaphragm supported by shear walls, what is load share at wall at axis C for a diaphragm force of 50 *kip* acting in the North direction ignoring the lateral stiffness of columns at line B?

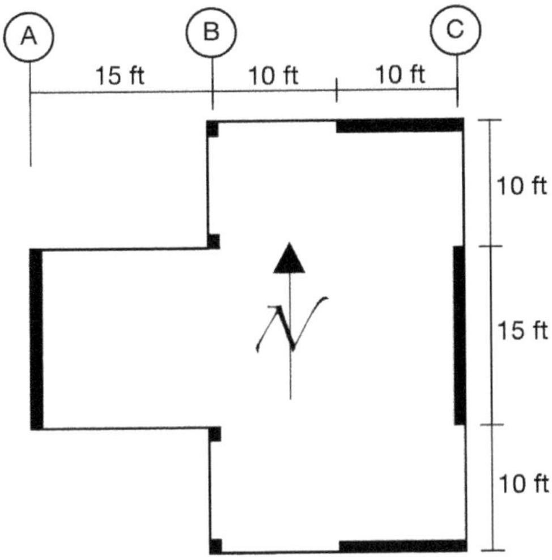

(A) 14.2 *kip*

(B) 18.9 *kip*

(C) 22.9 *kip*

(D) 29.6 *kip*

PROBLEM 1.38

The diaphragm below is supported by four shear walls with equal rigidity (assume rigidity of $R = 1$ per wall). The Center of Rigidity (CR) and the Center of Mass are both at the diaphragm's centroid.

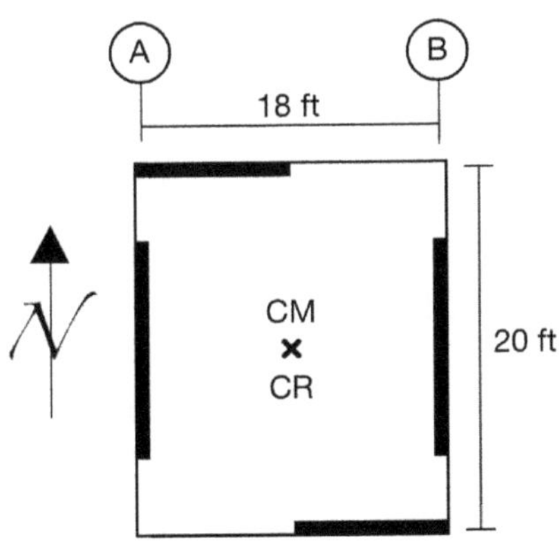

Given a 75 *kip* lateral load acting in the North direction, what is the maximum load share of shear wall A?

(A) 37.5 *kip*

(B) 39.2 *kip*

(C) 41.5 *kip*

(D) 43.5 *kip*

PROBLEM 1.39

Which of the following is the most effective and also feasible technique to improve the in-plane shear capacity of an inadequate unreinforced masonry wall?

(A) Add a thin surface treatment (e.g., plaster with wire mesh and cement mortar) at the inside and outside face of existing wall.

(B) Place reinforcing steel on the inside or outside face of the wall and apply new concrete layer.

(C) Add reinforcing steel to the wall utilizing the center coring technique.

(D) Fill in existing window or door openings with reinforced concrete or masonry.

PROBLEM 1.40

The walls below are made from Douglas-Fir-Larch with normal sheathing panels provided.

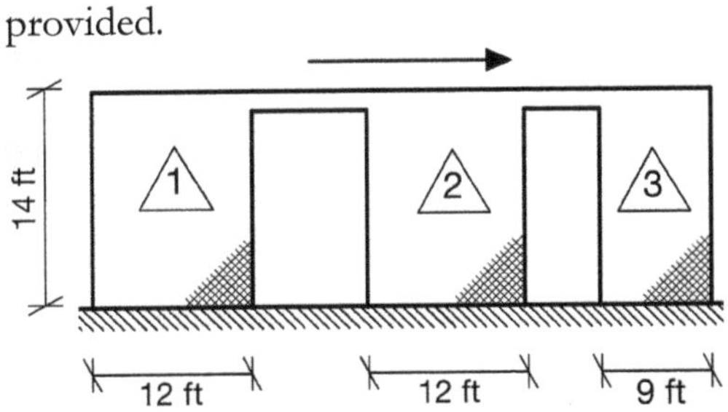

Wall	Sheathing thickness (*in*)	Nail bearing length in framing (*in*)	Nail type	Nail spacing (*in*)
1	5/16	1 − 1/4	6*d*	6 *in*
2	19/32	1 − 1/2	10*d*	6 *in*
3	7/16	1 − 3/8	8*d*	6 *in*

What is the nominal shear capacity for the three shear walls combined in the direction shown?

(A) 11.5 kip

(B) 18.1 kip

(C) 23.5 kip

(D) 29.8 kip

PROBLEM 1.41

What is the most common failure of traction elevators during earthquakes?

(A) Anchorage failure of electrical or mechanical equipment.

(B) Counterweights coming out of their guide.

(C) Damage to rail spreader beam brackets.

(D) Rail separation from structural supports.

PROBLEM 1.42

What factor should a wood diaphragm apparent shear stiffness (G_a) be adjusted with when moisture content reaches 21%?

(A) 0.5

(B) 0.6

(C) 0.7

(D) 0.85

PROBLEM 1.43

Structural observations are not required for which one of the following structures?

(A) Educational facility which provides IELTS and other language prep classes with occupant load of 300.

(B) A two-story residential building assigned to SDC E.

(C) The Engineer of Record (EOR) observation recommendation.

(D) A detention center.

PROBLEM 1.44

The sketch below shows a blocked wood diaphragm, with horizontal blocking strips as illustrated. The diaphragm uses standard layer sheathing for a single story, installed using two rows of $10d$ nails. The nails have a nominal length of 1 ½ in, spaced at 6 in on center throughout. The sheathing thickness is 19/32 in, and the width of the nailed face of the framing members adjoining the panels is 4 in.

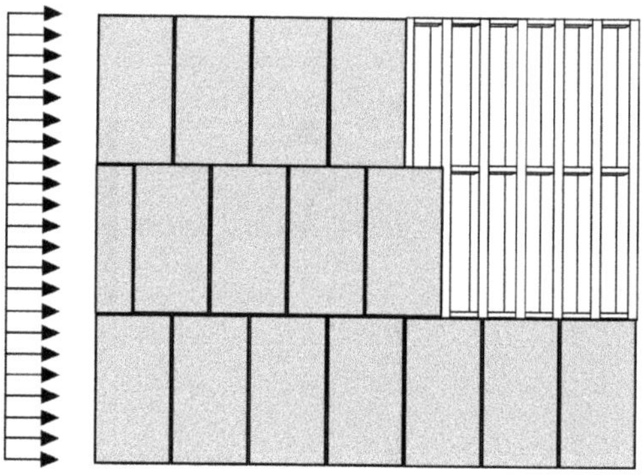

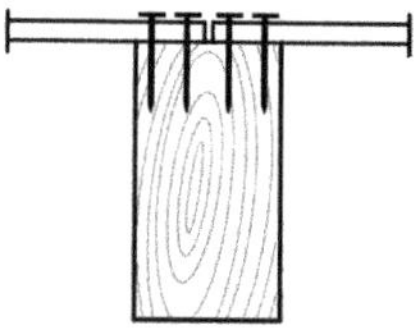

Based on this information, what is the allowable shear capacity in plf for the load acting in the direction indicated in the sketch if this diaphragm was made of Dougals-Fir Larch wood?

(A) 755 plf

(B) 936 plf

(C) 1,057 plf

(D) 2,115 plf

PROBLEM 1.45

A small skirt-supported pressure vessel, located on the second floor of an industrial reinforced concrete building, has a diameter of 36 in, a height of 9 ft, and a total weight (including the vessel and water) of 4.5 kip.

The centroid of mass is located at the center of the water vessel.

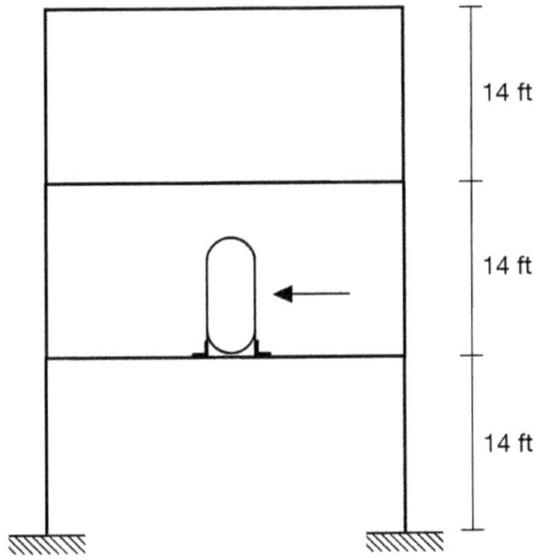

The seismic parameters for the building are as follows:

- $S_{DS} = 0.5$
- $S_{D1} = 0.3$
- Importance factor $I = 1.0$
- Building seismic weight $= 300\ kip$
- Building fundamental period $T = 0.5$

What is the moment generated at the base of the pressure vessel's support due to seismic excitation?

(A) $1.5\ kip.ft$

(B) $3.15\ kip.ft$

(C) $6.75\ kip.ft$

(D) $9.33\ kip.ft$

PROBLEM 1.46

A non-load-bearing partition wall is constructed using wood studs with gypsum board on both sides and has a unit weight of $12\ psf$. The wall is located on the first floor of a building and is $9\ ft$ tall. The site-specific seismic parameters are:

- $S_{DS} = 0.9$
- $S_{D1} = 0.7$
- $I_p = 1.5$

According to ASCE 7, what is the design seismic force acting on this wall?

(A) $23.3\ plf$

(B) $43.7\ plf$

(C) $58.3\ plf$

(D) $65.3\ plf$

PROBLEM 1.47

What is the minimum seismic design force that should be considered for a false/ suspended ceiling that has an operating weight of $2.5\ lb/ft^2$ located at the third floor of a five-story building, assigned to an SDC D with an importance factor of $I_p = 1.0$ and a seismic site coefficient of $S_{DS} = 0.55$?

(A) $0.22\ psf$

(B) $0.40\ psf$

(C) $0.66\ psf$

(D) $0.72\ psf$

PROBLEM 1.48

A $500\ ft$ tall chimney made of concrete and designed in accordance with ACI 307, is used in a power generating plant backing up water storage facilities, weighs $11{,}450\ kip$, has the following seismic parameters:

- $S_s = 1.22$
- $S_1 = 0.62$
- $T = 0.44\ sec$

Based on the above information, what is the minimum base shear that should be considered for this chimney?

(A) $2{,}634\ kip$

(B) $3{,}124\ kip$

(C) $3{,}334\ kip$

(D) $3{,}435\ kip$

PROBLEM 1.49

Which of the following nonstructural components is not exempt from seismic design requirements?

(A) A non-load-bearing gypsum partition wall in a low-rise commercial building in SDC B.

(B) An air conditioning unit installed on the roof of a commercial building in SDC B.

(C) A $300\ lb$ electrical panel attached with center of mass $3.5\ ft$ above the floor in a warehouse in SDC D, with flexible conduits and $I_p = 1.0$.

(D) A $75\ lb$ electrical panel attached with center of mass $4.5\ ft$ above the floor in a warehouse in SDC D, with flexible conduits and $I_p = 1.0$.

PROBLEM 1.50

A $30\ ft$ tall monument is laterally supported by a $30\ ft$ tall steel ordinary moment frame.

The monument has an effective weight of $5\ kip$ and the supporting structure of $10\ kip$.

The fundamental period of the monument is $0.05\ sec$, while that of the supporting moment frame is $0.26\ sec$.

Based on this information, what is the best method for seismically designing the supporting structure?

(A) The supporting structure to be designed using ASCE 7 Chapter 12 or 15.5 using $R = 3.5$.

(B) The supporting structure to be designed using ASCE 7 Chapter 12 or 15.5 using $R = 5$.

(C) The supporting structure and the monument to be modelled together using $R = 3.5$.

(D) The moment frame and the monument to be modelled together using $R = 5.5$.

PROBLEM 1.51

A $160\ ft$ tall, elevated equipment platform in an industrial location that uses a steel ordinary moment frame for an SFRS, assigned to an SDC D, with an effective seismic weight of $120\ kip$, has the following design parameters:

- $S_s = 0.88$
- $S_1 = 0.65$
- $T = 1.60\ sec$

With a Site Class D, risk category II, what is the base shear that should be considered for this nonbuilding structure?

(A) $54.0\ kip$

(B) $62.4\ kip$

(C) $71.3\ kip$

(D) $76.5\ kip$

PROBLEM 1.52

Which of the following statement(s) about impulsive and convective loads in tanks is TRUE (select all the true ones based on the options presented below)?

I. The convective response can govern the design of the tank roof.

II. The total lateral load on a tank is typically divided into impulsive and convective components.

III. The convective mode has a shorter natural period compared to the impulsive mode.

IV. Impulsive loads tend to be more significant in smaller tanks, whereas convective become more dominant in larger tanks.

(A) I

(B) I, II

(C) I, II, III

(D) I, II, IV

PROBLEM 1.53

Which of those statements regarding Reduced Beam Sections (RBS) is incorrect?

(A) In an RBS, the column is protected by reducing the moment transferred to it.

(B) An RBS cannot be placed in the protected zone.

(C) The location of the RBS is chosen near the beam-column connection to encourage plastic hinges to form at that location.

(D) An RBS is used in special moment frames and increases their ductility.

PROBLEM 1.54

Which of the following statements about the role of transverse reinforcement in columns of special moment frames, as required for seismic detailing, is incorrect?

(A) Confine the core concrete within those reinforcements.

(B) Provide lateral support to columns' longitudinal reinforcements.

(C) Confine longitudinal reinforcements at lap splice's locations.

(D) Improve the moment capacity of columns.

PROBLEM 1.55

What is the critical damping for a building in the context of seismic design?

(A) The damping level at which the building vibrates indefinitely.

(B) The damping level at which the building oscillates rapidly before settling.

(C) The damping value that prevents the building from oscillating.

(D) The damping level, at which, resonance takes place.

BONUS QUESTION

According to the 2022 California Building Code, how many stories are permitted for wooden light frames construction in Seismic Design Category D & E?

(A) One

(B) Two

(C) Three

(D) Four

Answer Key

No.	Our Answer	Your Notes
1.1	C	
1.2	A	
1.3	C	
1.4	B	
1.5	B	
1.6	C	
1.7	C	
1.8	A	
1.9	A	
1.10	C	
1.11	A	
1.12	B	
1.13	C	
1.14	D	
1.15	C	
1.16	D	
1.17	A	
1.18	A	
1.19	A	
1.20	D	
1.21	B	
1.22	B	
1.23	C	
1.24	C	
1.25	A	
1.26	A	
1.27	C, D, E	
1.28	B	
1.29	A, C, E	
1.30	A	
1.31	B	
1.32	A	
1.33	C	
1.34	A	
1.35	D	
1.36	B	
1.37	D	
1.38	B	
1.39	B	
1.40	C	
1.41	B	
1.42	A	
1.43	B	
1.44	A	
1.45	C	
1.46	B	
1.47	C	
1.48	A	
1.49	D	
1.50	A	
1.51	B	
1.52	D	
1.53	B	
1.54	D	
1.55	C	

SOLUTION 1.1

The following equation, in *The Essential Guide* Section 1.4.1 along with Table 1-1, is applied to the first 100 ft:

$$\overline{S_u} = \frac{\sum_{i=1}^{k} d_i}{\sum_{i=1}^{k} \frac{d_i}{s_{ui}}}$$

Where is d_i the layer thickness and k refers to the cohesive layers.

$$\overline{s_u} = \frac{35 + 65}{\frac{35}{1{,}400} + \frac{65}{800}} = 941.1\ psf$$

In reference to Table 1-1 of *The Essential Guide* with $\overline{S_u} = 941.1\ psf$, this site is classified as **Site Class E – Soft Clay Soil**.

The above information can be found in the ASCE 7 Standard, Section 20.4 Definitions of Site Class Parameters, and Table 20.3-1 Site Classification.

Correct Answer is (C)

SOLUTION 1.2

The correct answer is the **marine clay**. Marine clay is a cohesive soil with low permeability and high compaction, making it highly resistant to liquefaction.

As for the other options, they have higher potential for liquefaction as expanded below:

Coastal Reclaimed Land and Artificial Fill
These are typically composed of loose, poorly compacted materials, often water-saturated due to proximity to water bodies. Their loose structure and high-water content makes them prone to liquefaction.

Soft, Unconsolidated Marine Deposits
These are loose sands and silts near shorelines, are highly susceptible to liquefaction. Their lack of compaction and water saturation makes them prone to liquefaction under seismic shaking.

Loose, Saturated Sands and Silts in Coastal Areas
The combination of loose grain structure and high-water content leads to loss of strength during an earthquake, allowing these soils to behave like a liquid.

For more information on liquefaction, see *The Essential Guide* Section 1.4.3.

Correct Answer is (A)

SOLUTION 1.3

Referring to page 5 of *The Essential Guide* for this cross section, where it can be concluded from the average shear velocity, with a thick layer of bedrock, $\bar{v}_s > 2{,}500\ ft/s$, this site could be classified as Class B.

However, the same section states as well that if the there is more than 10 ft of soil between the rock surface and the bottom of the spread footing then it cannot be assigned to A or B, and hence, a **Site Class C** would be the most appropriate in this case.

The above information can be found in the ASCE 7 Standard, Section 20.1.

Correct Answer is (C)

SOLUTION 1.4

Given no site information are known; the Site Class can be assumed as D and $F_a > 1.2$. Based on this, and consulting with Tables 1-2 and 1-3 of *The Essential Guide* to determine values of F_a and F_v:

$$S_{DS} = \frac{2}{3} \times F_a \times S_s = \frac{2}{3} \times 1.6 \times 0.25 = 0.27$$

$$S_{D1} = \frac{2}{3} \times F_v \times S_1 = \frac{2}{3} \times 2.4 \times 0.10 = 0.16$$

The risk category for a police station is IV based on Section 6.2.2 page 210 of *The Essential Guide*.

Using Table 1-7 of *The Essential Guide* with the above information generates an SDC C, while using Table 1-8 gives us an SDC D.

Based on the information provided in the ASCE and the same section of *The Essential Guide*, the more severe SDC prevails, in which case **SDC D**.

> Information provided in this solution can be found in the following sections of codes:
>
> - F_a & F_v: ASCE 7 Table 11.4-1 & 2.
> - Information on SDC D when soil information is not known: ASCE Section 11.4.3 & 11.4.4.
> - SDC Tables: ASCE Tables 11.6-1 & 2.
> - Risk categories can be found in CBC 2022 Section 1604.5.

Correct Answer is (B)

SOLUTION 1.5

We need to find the SDC for this building referring to the methods used in the previous question, then use SDC to compare the height limitations for each SFRS in ASCE 7 Table 12.2-1.

Given a Site Class C and consulting with Tables 1-2 and 1-3 of *The Essential Guide* to determine values of F_a and F_v:

$$S_{DS} = \frac{2}{3} \times F_a \times S_s = \frac{2}{3} \times 1.2 \times 0.75 = 0.60$$

$$S_{D1} = \frac{2}{3} \times F_v \times S_1 = \frac{2}{3} \times 1.5 \times 0.25 = 0.25$$

The risk category for a building with public assemblies and a total occupant load less than 2,500 is II. This is based on Section 6.2.2 page 209 of *The Essential Guide*.

Using Table 1-7 & 1-8 of *The Essential Guide* with the above information generates an SDC D.

Based on the above information, with a building height of 65 ft and SDC D, the best system from the above selections is the **light wood frame walls sheathed with wood structural panels rated for shear resistance.**

> Information provided in this solution can be found in the following sections of codes:
>
> - F_a & F_v: ASCE 7 Table 11.4-1 & 2.
> - SDC Tables: ASCE Tables 11.6-1 & 2.
> - Risk categories can be found in CBC 2022 Section 1604.5.

Correct Answer is (B)

SOLUTION 1.6

According to Table 12.2-1 of ASCE 7, the most appropriate seismic force-resisting system for this scenario is **light-frame (cold-formed steel) walls sheathed with wood structural panels rated for shear resistance**.

While Option A, which involves special reinforced concrete shear walls, may be technically acceptable, it is not the most suitable choice for this type of building and would likely lead to a less economical and more complex solution.

Correct Answer is (C)

SOLUTION 1.7

Use Table 3-1 of *The Essential Guide* and Table 12.8-1 of ASCE 7 to determine the approximate period as follows:

$$T_a = C_t h_n^x$$
$$= 0.028 \times (14 \times 12)^{0.8}$$
$$= 1.7\ sec$$

Frequency is the inverse of the fundamental period – see the notes section in *The Essential Guide.*

$$F = \frac{1}{T} = \frac{1}{1.7} = 0.6\ Hz$$

The above information on fundamental period is in ASCE 7 Section 12.8.2.1.

Correct Answer is (C)

SOLUTION 1.8

Refer to Section 3.2.4 of *The Essential Guide* to use the fundamental period equation and invert it to get the natural frequency as follows:

$$T = 2\pi\sqrt{\frac{W}{kg}}$$

$$\rightarrow f = \frac{1}{T} = \frac{1}{2\pi\sqrt{\frac{W}{kg}}}$$

When all coefficients of this equation remain unchanged, we can compare the new frequency with the current one using substitution, as follows:

$$f_{current} = \frac{1}{2\pi\sqrt{\frac{W}{kg}}}$$

$$f_{new} = \frac{1}{2\pi\sqrt{\frac{W}{2kg}}} = \frac{\sqrt{2}}{2\pi\sqrt{\frac{W}{kg}}} = \sqrt{2} \times f_{current}$$

Information on the calculation of the fundamental period can be found in ASCE 7, Section 13.3.3. The same equation provided for components in that section can also be applied to single-degree-of-freedom systems.

Correct Answer is (A)

SOLUTION 1.9

By increasing the amount of reinforcement and improving its spacing, an ordinary reinforced masonry shear wall can be upgraded to either an intermediate or special reinforced masonry shear wall.

According to ASCE 7 Table 12.2-1, the response modification factor (R), which reflects the ductility and energy dissipation capacity of the system, increases when using intermediate or special reinforced masonry walls compared to ordinary ones.

For a summary of the minimum reinforcement requirements for ordinary, intermediate, and special reinforced masonry walls, refer to Figures 6-5 through 6-8 in *The Essential Guide*, as well as Section 7.3 of TMS 402/602-16.

Information on reinforcement for masonry shear wall can be found in TMS 402/602 Section 7.3.

Correct Answer is (A)

SOLUTION 1.10

Reference in this question is made to *The Essential Guide* Section 2.4.1 and 2.4.3 (for setbacks) and Section 6.2.2 page 209 (for risk categories), along with ASCE 7 Table 12.2-1.

Determine the deflection amplification factor (C_d) for each building, after which, determine the maximum inelastic response at floor 3 only for the two buildings as this is the floor where pounding is most likely to occur.

Coefficient	Building 1	Building 2
C_d	2.5	4.5
Risk Category	II	III
I_e	1.0	1.25
$\delta_{e\ @\ level\ 3}$	3.5 in	3.7 in

$$\delta_{M1} = \frac{C_d \delta_{max}}{I_e}$$
$$= \frac{2.5 \times 3.5\ in}{1.0}$$
$$= 8.75\ in$$

$$\delta_{M2} = \frac{C_d \delta_{max}}{I_e}$$
$$= \frac{4.5 \times 3.7\ in}{1.25}$$
$$= 13.32\ in$$

$$\delta_{MT} = \sqrt{(\delta_{M1})^2 + (\delta_{M2})^2}$$
$$= \sqrt{(8.75)^2 + (13.32)^2}$$
$$= 15.94\ in$$

Information on buildings' deflection and setbacks are in ASCE 7 Section 12.12.3.

Risk categories can be found in CBC 2022 Section 1604.5.

Correct Answer is (C)

SOLUTION 1.11

Per The Essential Guie Table 2-1, accidental torsional moment and their amplification factors are only used for rigid and semi-rigid diaphragms → *zero.*

Information on this can be found in ASCE 7 Section 12.8.4.3 and Table 12.3-1 items 1a and 1b.

Correct Answer is (A)

SOLUTION 1.12

Table 2-2 of *The Essential Guide* states that for Type 5.b lateral strength irregularity, a weak story is permitted to have a maximum height of 30 ft or two stories without the need to apply the overstrength factor.

This information can be found in ASCE 7 Section 12.3.3.2.

Correct Answer is (B)

SOLUTION 1.13

ASCE 7 Section 12.8.3 uses a simplified method to vertically distribute base shear forces for the height of the structure. This method assumes the structure's response is controlled by the first (fundamental) mode of vibration, especially for regular, low- to mid-rise buildings.

Correct Answer is (C)

SOLUTION 1.14

This condition is most commonly observed in small wood-frame buildings, where discontinuities in diaphragms or weak interconnections can lead to separation of building parts during seismic shaking. According to Table 7-2 of *The Essential Guide*, this behavior is far more prevalent in light-frame wood construction than in other structural systems listed.

This information can also be found in ATC 20, page 35, point 1.

Correct Answer is (D)

SOLUTION 1.15

Reference is made here to ASCE 7 Section C12.11.2.2.4. In this case, in order to manage diaphragm forces perpendicular to the deck corrugations, it is common for metal decks to be supported at 8 ft to 10 ft intervals by joists that are connected to walls in a manner suitable to resist the full wall anchorage design force and to carry that force across the diaphragm.

Correct Answer is (C)

SOLUTION 1.16

Reference is made to *The Essential Guide*, Table 1-2 to determine the required site coefficients, page 210 to determine risk category which is IV, and with the use of

Table 1-5 in this case $I_e = 1.5$, and finally Section 3.3.5 to determine the anchorage forces.

With a Site Class C and $S_s = 0.9 \rightarrow F_a = 1.2$:

$$S_{DS} = \frac{2}{3} \times 0.9 \times 1.2 = 0.72$$

The tributary area is shown in Figure 3-5 of *The Essential Guide* and is calculated per linear width of the wall as follows:

$$A = 1\,ft \times \left(\frac{12\,ft}{2} + 4\,ft\right) = 10\,ft^2/ft$$

$$W_p = \left(150\frac{lb}{ft^3}\right) \times \left(\frac{9}{12}\,ft\right) \times \left(10\frac{ft^2}{ft}\right)$$

$$= 1{,}125\,plf$$

Calculate the amplification factor for flexile diaphragms:

$$k_a = 1 + \frac{L_f}{100}$$

$$= 1 + \frac{80\,ft}{100} = 1.8 < 2.0$$

$$F_p = 0.4 S_{DS} k_a I_e W_p$$

$$= 0.4 \times 0.72 \times 1.8 \times 1.5 \times 1{,}125\,plf$$

$$= 874.8\,plf\ (*)$$

Check against the minimum force as follows:

$$F_{p,min} = 0.2 k_a I_e W_p$$

$$= 0.2 \times 1.8 \times 1.5 \times 1{,}125\,plf$$

$$= 607.5\,plf < 874.8\,plf \rightarrow OK$$

The information mentioned above can be found in the following codes:

- F_a: ASCE 7 Table 11.4-1.
- Risk categories can be found in CBC 2022 Section 1604.5.
- Anchorage forces: ASCE 7 Section 12.11.2.1.
- Discussion about tributary area for this case is in ASCE 7 C12.11.2.2.7.

Correct Answer is (D)

(*) There is an alternative approach used by some, where the parapet is excluded from the calculation of the tributary area and the wall weight W_p and added later. In this method, the parapet force is calculated independently using the guidelines from Chapter 13 of ASCE 7 (Section 5.3.3 of *The Essential Guide*), then added to the force generated by the wall alone. This method results in a higher anchorage force. However, based on my experience, I have not found this method to be the one commonly used in the Seismic Principles Exam. Instead, the approach presented here is typically the one expected to be followed.

SOLUTION 1.17

Reference is made to *The Essential Guide*, Section 3.2.2 for the effective weight calculation.

The stored goods are less than 5% of the store effective weight:

$$\frac{7.5}{175} = 4.3\% < 5\%$$

Based on this, we do not need to include the live load into the calculation of the effective seismic weight → *zero*.

The information mentioned above can be found in the following codes:

- Effective seismic weight: ASCE 7 Section 12.7.2.

Correct Answer is (A)

SOLUTION 1.18

Reference is made to *The Essential Guide*, Table 2-1 which tells you that the redundancy factor $\rho = 1.3$ for type 1b horizontal irregularities should be used, risk categories page 210 classifying this building as risk category IV, and finally to ASCE 7 Table12.12-1 located in Chapter 2 appendix of *The Essential Guide*.

Note *b* under ASCE 7 Table 12.12-1 provides the following recommendation for moment-resisting frames structures:

$$\Delta_{design} < \frac{\Delta_a}{\rho}$$

$$< \frac{0.01h_{sx}}{1.3}$$

$$< \frac{0.01 \times 12\,ft \times 12\frac{in}{ft}}{1.3}$$

$$< 1.1\,in$$

The information mentioned above can be found in the following codes:

- Design treatments for horizontal irregularities: ASCE 7 Table 12.3-1.
- Value of $\rho = 1.3$: ASCE 7 Section 12.3.4.2.
- Risk categories can be found in CBC 2022 Section 1604.5.
- Allowable story drifts: ASCE 7 Section 12.12.

Correct Answer is (A)

SOLUTION 1.19

The correct cross-section is the one that closely resembles the cross-section shown in *The Essential Guide*, Figure 3-7 on page 95, which is cross-section of Option A.

Additionally, it is important to note that according to SDPWS 2021 Section 4.1.5.1.1, toenails are not permitted for anchoring, which is the one shown in Option C.

It is advisable to explore reliable online resources or consult the reference list provided in *The Essential Guide* for more anchorage cross-sections as this is a common topic.

Correct Answer is (A)

SOLUTION 1.20

Reference is made to *The Essential Guide*, Section 3.4.1. This section states that the most stringent condition should be used in this case, which represents the lowest response modification factor of 1.5 for the plain AAC masonry shear walls.

The information mentioned above can be found in the following codes:

- SFRS horizontal combination: ASCE 7 Section 12.2.3.3.
- ASCE 7 Table 12.2-1 for the various SFRS system coefficients.

Correct Answer is (D)

SOLUTION 1.21

Since rigidity for each wall is given in this question, calculating the center of rigidity is straight forward and is as follows:

$$\bar{x} = \frac{0 \times 1 + 20 \times 3}{1 + 3} = 15$$

$$\bar{y} = \frac{0 \times 2 + 15 \times 2}{2 + 2} = 7.5$$

Correct Answer is (B)

SOLUTION 1.22

Check the aspect ratio for the longest wall:

$$\frac{h}{l} = \frac{16}{4} = 4 > 3$$

The above indicates that the walls are slender and that they will experience flexural deflection (neither shear deflection nor combined as explained in Section 3.2.3 of *The Essential Guide*).

The Essential Guide Table 4-1 explains that the moment of inertia for each wall can be used to determine the center of rigidity especially that all walls have the same height, and their end conditions do not differ:

There are two wall types here, 4 ft long and 3 ft long:

$$I_{4ft} = \frac{1\,ft \times (4\,ft)^3}{12} = 5.33\,ft^4$$

$$I_{3ft} = \frac{1\,ft \times (3\,ft)^3}{12} = 2.25\,ft^4$$

$$\bar{x} = \frac{0 + 11 \times 5.33 + 22 \times (5.33 + 2.25)}{5.33 + 5.33 + 5.33 + 2.25} = 12.4$$

$$\bar{y} = \frac{0 \times 5.33 + 10 \times (5.33 + 5.33)}{5.33 + 5.33 + 5.33} = 6.7$$

Correct Answer is (B)

SOLUTION 1.23 (*)

Using *The Essential Guide* Section 3.3.2, 3 & 4 and Table 3-4, along with Table 12.2-1 of the ASCE 7 to collect the response modification factor of the mentioned SFRS as $R = 6$, see below:

$$T_s = \frac{S_{D1}}{S_{DS}} = \frac{0.8}{0.6} = 1.33 > T\,(= 0.45)$$

$$\rightarrow C_s = \frac{S_{DS}}{\left(\frac{R}{I_e}\right)} = \frac{0.6}{\left(\frac{6}{1}\right)} = 0.1$$

$$C_{s,min} = 0.044 S_{DS} I_e = 0.044 \times 0.6 \times 1 = 0.026$$

$$V = C_s W = 0.1 \times (4 \times 200 + 125) = 92.5\,kip$$

Use the equation presented in Section 3.3.3 for vertical distribution as follows:

$$F_x = C_{vx} V$$

$$C_{vx} = \frac{w_x h_x^k}{\sum_{i=1}^{n} w_i h_i^k} \quad \text{where } k = 1 \text{ for } T < 0.5$$

	h_x	h_x^k	w_x	$w_x h_x^k$	C_{vx}	F_x
5	60	60	125	7,500.0	0.24	22.2
4	48	48	200	9,600.0	0.30	27.7
3	36	36	200	7,200.0	0.23	21.3
2	24	24	200	4,800.0	0.15	13.9
1	12	12	200	2,400.0	0.08	7.4
				31,500.0		**92.5**

Now, calculate the diaphragm design force at the third level as follows:

$$F_{px} = \frac{\sum_{i=3}^{5} F_i}{\sum_{i=3}^{5} w_i} w_{px}$$

$$= \frac{21.3 + 27.7 + 22.2}{200 + 200 + 125} \times 200 = 27.1\,kip$$

Check for minimum and maximum diaphragm forces:

$$F_{px,min} = 0.2 S_{DS} I_e w_{px}$$
$$= 0.2 \times 0.6 \times 1 \times 200$$
$$= 24\,kip$$

$$F_{px,min} = 0.4 S_{DS} I_e w_{px}$$
$$= 0.4 \times 0.6 \times 1 \times 200$$
$$= 48\,kip$$

→ Use $F_{px} = 27.1\,kip$

The information mentioned above can be found in ASCE 7 Sections:

- C_s: Section 12.8.1.1 Equation 12.8-3 and Section C12.8.1.1 for comparison with T_s.
- C_{vx}: Section 12.8.3.
- F_{px}: Section 12.10.1.

Correct Answer is (C)

(*) This question is longer than the other questions and may take up to 5 minutes to solve, involving several steps. Expect a few questions like this.

SOLUTION 1.24

Use *The Essential Guide* Table 3-1 to calculate the fundamental period along with ASCE 7 Table 12.8-2.

$$T_a = C_t h_n^t = 0.02 \times 120^{0.75} = 0.73\,sec$$

$$S_{DS} = \frac{2}{3} \times F_a \times S_s = \frac{2}{3} \times 0.9 \times 0.35 = 0.21$$

$$S_{D1} = \frac{2}{3} \times F_v \times S_1 = \frac{2}{3} \times 0.8 \times 0.65 = 0.35$$

$$T_s = \frac{S_{D1}}{S_{DS}} = \frac{0.35}{0.21} = 1.67 > T\ (= 0.73)$$

Moreover, use Table 12.2-1 of the ASCE 7 to collect the response modification factor of the mentioned SFRS as $R = 5$.

Using Table 3-4 of *The Essential Guide*:

$$\rightarrow C_s = \frac{S_{DS}}{\left(\frac{R}{I_e}\right)} = \frac{0.21}{\left(\frac{5}{1}\right)} = 0.042$$

Check $C_{s,min}$ using the following equation because $S_1 > 0.6$:

$$C_{s,min} = \frac{0.5S_1}{\left(\frac{R}{I_e}\right)} = \frac{0.5 \times 0.65}{\left(\frac{5}{1}\right)} = 0.065$$

→Use $C_s = 0.065$

The information mentioned above can be found in ASCE 7 Sections:

- S_{DS} & S_{D1}: Section 11.4.5.
- T_a: Section 12.8.2.1.
- C_s: Section 12.8.1.1 Equation 12.8-3 and Section C12.8.1.1 for comparison with T_s.
- $C_{s,min}$: Equation 12.8-6 when $S_1 \geq 0.6g$.

Correct Answer is (C)

SOLUTION 1.25

The Essential Guide Figure 1-13 shows that overhangs for cantilevers for the simplified method should not exceed fifth the width of the overhand for "other than cast-in-place concrete", and hence the maximum is $30/5 = 6\ ft$.

The information mentioned above can be found in ASCE 7 Section 12.14.1.1.

Correct Answer is (A)

SOLUTION 1.26

Reference is made to Section 5.4.6 of *The Essential Guide*. See below

$$T = 2\pi\sqrt{\frac{\sum w_i\delta_i^2}{g\sum f_i\delta_i}}$$

$$= 2\pi\sqrt{\frac{73\times1.7^2+73\times2.2^2}{32.17\frac{ft}{sec^2}\left(\times\frac{12\ in}{1\ ft}\right)\times(5\times1.7+17\times2.2)}}$$

$$= 1.12\ sec$$

The information mentioned above can be found in ASCE 7 Section 15.4.4.

Correct Answer is (A)

SOLUTION 1.27

When answering these types of questions, it is important to visualize how lateral loads from the upper levels transfer to lower levels.

For wood-framed structures, the SDPWS discourages relying on nail withdrawal to resist shear, anchorage, or similar lateral forces or axial tension forces. This makes **Options A & B** ineffective, as they depend primarily on nail withdrawal rather than direct shear transfer especially if vertical forces from seismic or wind uplift were involved.

In those two options, it can be clearly seen that the top shear wall is only nailed to the diaphragm joist, which can be detached during seismic excitation (both lateral and vertical), notwithstanding any uplift from wind forces.

Effective lateral load paths typically involve sheathing panels sharing a common framing element, which ensures continuity in force transfer. This can be a joist, a framing element, or blocking. See below:

- **Option C** shows that the diaphragm and wall sheathing are attached to a shared top plate or framing member, allowing for a continuous load path.

- **Option D** shows the sheathing nailed into the diaphragm joist, which, despite being parallel to the wall, still functions as a shared element.

- **Option E** utilizes both top and bottom blocking as common members between the wood panels shown, effectively tying the system together.

In contrast, **Option F** is ineffective. The diaphragm and wall sheathing are each attached to separate framing members that are not interconnected. Even if those framing elements were connected, they would not act as a true collector.

For proper seismic load path continuity, the sheathing must share a common, continuous framing member – could be referred to as a collector – to transfer the forces effectively.

Correct Answer is (C, D & E)

SOLUTION 1.28

For this question, refer to *The Essential Guide*, Section 4.5.1, which presents two LRFD load combination equations that include seismic loads (load combination 6 and 7). These can be applied here. We will evaluate both equations to determine which one aligns with the options provided.

Additionally, as noted in *The Essential Guide*, page 148, item 2 – and also in Table 2-1 of Chapter 2 – overstrength (Ω_0) must be included in the load combination for this element due to the presence of an out-of-plane irregularity (see ASCE 7 Section 12.3.3.3). The overstrength factor for the specified SFRS is $\Omega_0 = 3.0$. Since overstrength is required, E_{mh} will be used for the horizontal seismic effects.

Load combination 6 per *The Essential Guide* page 147 (ASCE 7 Section 2.3.6):

$$1.2D + E_v + E_{mh} + L + 0.2S$$

Where:

$$E_v = 0.2S_{DS}D = 0.2 \times 0.5D = 0.1D$$

$$E_{mh} = \Omega_o Q_E = 3Q_E$$

$$L = L \,\&\, S = 0$$

The final equation is therefore:

$$1.2D + E_v + E_{mh} + L + 0.2S$$
$$= 1.2D + 0.1D + 3Q_E + L + 0$$
$$\rightarrow 1.3D + 3Q_E + L$$

Load combination 7:

$$0.9D - E_v + E_h \ (or\ E_{mh})$$

The final equation is therefore:

$$0.8D + 3Q_E$$

Based on this, Option B is similar to the final load combination 7.

> The information mentioned above can be found in ASCE 7 Sections 2.3.6, 12.4 and 12.3.3.3.

Correct Answer is (B)

SOLUTION 1.29

The Essential Guide page 148 provides a list, based upon which, **Options A, C & E** are the only conditions from this list where overstrength shall be applied.

> The information mentioned above can be found in the following ASCE 7 Sections:
>
> **Option A:** Section 12.2.5.2 where overstrength must be used in foundation and overturning elements at the base of cantilever columns system.

Option C: Section 12.10.2.1 where collector elements in SDC C through F require overstrength.

Option E: Section 12.13.8.4 where batter piles and their connections must be designed using overstrength.

Correct Answer is (A, C & E)

SOLUTION 1.30

Equations for flexible diaphragms can be taken from Figure 4-2 of The Essential Guide as follows:

$$M_{chord\ @\ 9ft} = \frac{wx(L-x)}{2}$$

$$= \frac{200\frac{lb}{ft}\times 9ft\times(36-9)\ ft}{2}$$

$$= 24{,}300\ lb.ft$$

$$T_{chord\ @\ 9ft} = \frac{M_{chord\ @\ 9ft}}{D}$$

$$= \frac{24{,}300\ lb.ft}{30\ ft}$$

$$= 810\ lb\ (0.81\ kip)$$

Correct Answer is (A)

SOLUTION 1.31

The double top plates in the wall, as shown in **Option B**, are typically used as the chord of the diaphragm. To be more accurate, it is normally the top one.

In this case, the two top plates are spliced – i.e., their ends are staggered rather than aligned.

Option A represents blocking for the main frame. The blocking may be discontinuous and therefore cannot be relied on to resist tension forces.

Option C points to an L-section added above the shear wall two top plates to connect the diaphragm to the wall, and hence this does not act as a chord.

Option D represents the continuation of the sheathing, which helps tighten the overall structure but does not serve as the chord for the diaphragm.

Correct Answer is (B)

SOLUTION 1.32

This case is illustrated in Figure 4-4 of *The Essential Guide*. In such situations, the chord tension reinforcement cannot be integrated with the beam reinforcement, as it may interfere with the design of the tension rebars and the compression block. Therefore, the best placement is as shown in **Option A**.

Additionally, these reinforcements can also serve as part of the shrinkage reinforcement of the slab.

Correct Answer is (A)

SOLUTION 1.33

Check the aspect ratio for the two walls:

$$\frac{h_{left}}{l_{left}} = \frac{12}{15} = 0.8 < 3$$

$$\frac{h_{right}}{l_{right}} = \frac{12}{12} = 1.0 < 3$$

The above indicates that the walls will experience a combined flexural and shear deflection as explained in Section 3.2.3 of *The Essential Guide*.

The Essential Guide Table 4-1 explains that we shall use the rigidities provided in Table 3-6 for this purpose, and for cantilever walls – see the below grey text box if you did not own a copy of *The Essential Guide* for a long solution (*).

$$R_{left} = 0.2248$$

$$R_{right} = 0.1429$$

$$V_{left} = \frac{0.2248}{0.2248 + 0.1429} = 15.3\ kip$$

The following equations outline how to calculate the above relative rigidities:

$$R_{left} = \frac{1}{4\left(\frac{h}{l}\right)^3 + 3\left(\frac{h}{l}\right)} = \frac{1}{4(0.8)^3 + 3(0.8)} = 0.2248$$

$$R_{right} = \frac{1}{4\left(\frac{h}{l}\right)^3 + 3\left(\frac{h}{l}\right)} = \frac{1}{4(1.0)^3 + 3(1.0)} = 0.1429$$

Correct Answer is (C)

(*) You may notice that in some resources, these values are presented as 10 times larger to emphasize rigidity. This is often referred to as relative rigidity. Both approaches are valid as long as the same method is used consistently throughout the solution. However, the values provided here are more accurate.

SOLUTION 1.34

We will address this question step-by-step to explain the underlying theory.

Initially, it is important to note that this diaphragm is flexible. Consequently, the forces from the diaphragm's lateral shear are distributed according to the tributary area, or alternatively, utilizing a simple beam method.

$$V = \frac{200\ lb/ft \times 36\ ft}{2} = 3{,}600\ lb\ (3.6\ kip)$$

The unit shear taken by the shear wall alone is:

$$= \frac{3.6\ kip}{20\ ft} = 0.18\ klf$$

The unit shear taken by the entire length of the collector which covers the full 27 ft is:

$$= \frac{3.6\ kip}{27\ ft} = 0.1\overline{33}\ klf$$

The two unit forces mentioned above are acting in opposition to each other. For better understanding, they are plotted as follows, and the collector load diagram is shown below the wall elevation.

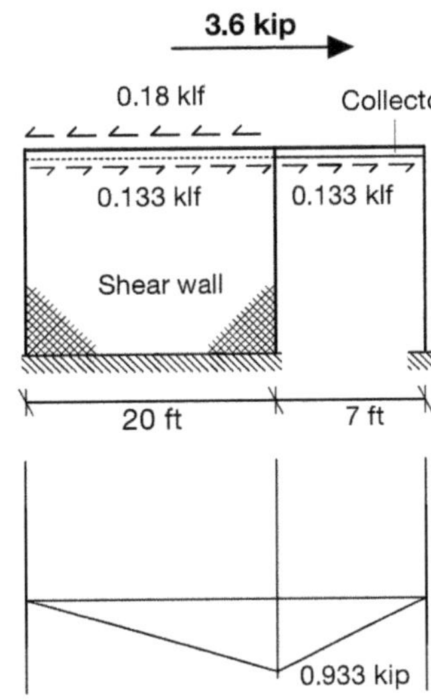

It is important to note that the force in the collector is in tension as it pulls the shear wall in the direction shown, hence we are showing it in negative on the diagram.

To construct the left side of the diagram, it is important not to forget subtracting the $0.1\overline{33}\ klf$ from $0.18\ klf$ prior to proceeding with the diagram construction.

Correct Answer is (A)

(*) More examples and theory explanation are available in *The Essential Guide.*

SOLUTION 1.35

In a similar way to the previous question, we will address this question step-by-step to explain the underlying theory.

This diaphragm is flexible, and hence, forces from the diaphragm's lateral shear are distributed according to the tributary area.

$$V = \frac{600\ lb/ft \times 22\ ft}{2} = 6{,}600\ lb\ (6.6\ kip)$$

The shear walls are composed of wood, as indicated by the question. Therefore, the shear can be distributed to these walls based on their length. This is explained in *The Essential Guide* diagram in Figure 4-1.

$$= \frac{6.6\,kip}{10\,ft + 15\,ft} = 0.264\,klf$$

The unit shear taken by the entire length of collector which covers the full 40 ft is:

$$= \frac{6.6\,kip}{40\,ft} = 0.165\,klf$$

The two unit forces mentioned above are acting in opposition to each other. For better understanding, they are plotted as follows, and the collector load diagram is shown below the wall elevation.

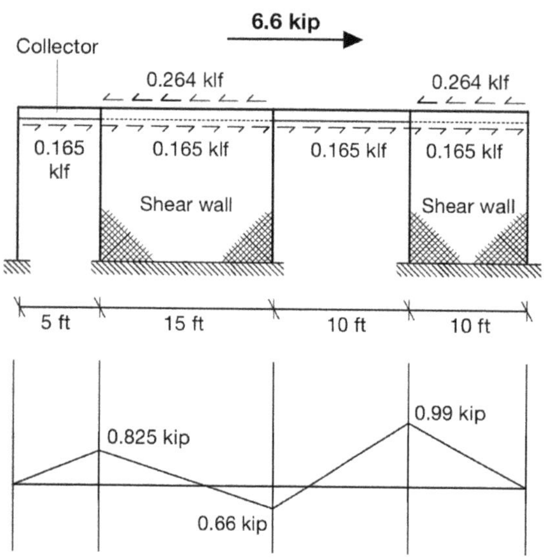

To construct the diagram at the shear wall locations, envision the forces being applied to the shear walls by the collectors, with the shear walls providing resistance along their entire line of defense. Therefore, it is important to determine the resultant force of both the collector unit load and the shear wall unit load as outlined below (*):

- For the 15 ft wall:

$$= 0.264 - 0.165 = 0.099\,klf$$

- For the 10 ft wall:

$$= 0.264 - 0.165 = 0.099\,klf$$

The maximum force is therefore 0.99 kip as shown on the diagram.

Correct Answer is (D)

(*) When constructing this diagram, it is essential to treat it as a force bookkeeping exercise. The diagram should start at zero from one side and closes at zero on the other side.

From the collector diagram, note that the collector on the left side is in compression as it exerts force on the shear wall (depicted as positive on the diagram), while the collector in the middle starts in tension as it pulls the left side shear wall, then switches to compression as it pushes the rightmost shear wall, transferring this force along its length from the attached diaphragm.

More examples and theory explanation are available in *The Essential Guide.*

SOLUTION 1.36

Since this is a flexible diaphragm, the lateral force of 50 kip can be disturbed over the shear wall lines of resistance using tributary area as follows:

$$Area\ A\ to\ B = 15 \times 15$$
$$= 225\,ft^2\ (24.3\%)$$

$$Area\ B\ to\ C = 20 \times 35$$
$$= 700\,ft^2\ (75.7\%)$$

$$R_{wall\ C} = \frac{50\,kip \times 75.7\%}{2} = 18.93\,kip$$

Correct Answer is (B)

SOLUTION 1.37

Given that the diaphragm is flexible and non-uniform between supports, it can be modeled as a simply supported beam spanning lines A and B, as explained in Figure 4-1 of *The Essential Guide.*

First, determine the center of the resultant load of 50 kip which would be located at the centroid of the diaphragm.

Take axis at line A as datum:

$$\bar{X} = \frac{7.5 \times (15 \times 15) + 25 \times (20 \times 35)}{15 \times 15 + 20 \times 35} = 20.7\,ft$$

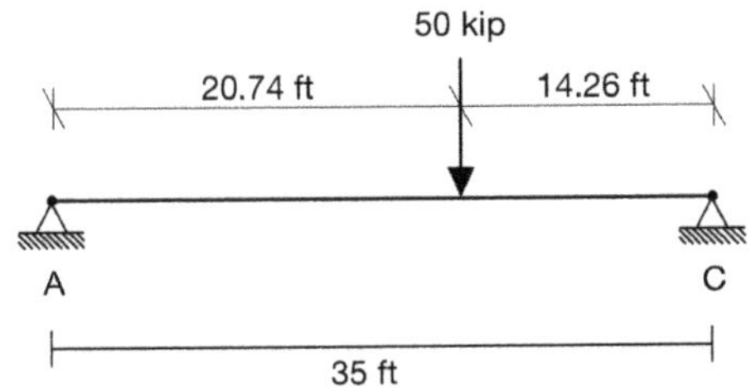

Take moment around support A to find reaction at support C:

$$R_{wall\ C} = \frac{50\ kip \times 20.74}{35} = 29.63\ kip$$

Correct Answer is (D)

SOLUTION 1.38

If you own a copy of *The Essential Guide*, check Section 4.3.3 for detailed explanation of the methodology used here.

ASCE 7 Section 12.8.4.2 requires that accidental torsion (M_{ta}) due to displacement of CM perpendicular to the direction of loading direction to be considered.

$$e_{CM\ displacment} = 5\% \times 18\ ft = \mp 0.9\ ft$$

$$M_{ta} = \mp 0.9 \times 75 = \mp 67.5\ kip.ft.$$

Using the North direction as the Y-axis, apply Equation 1a from Section 4.3.3 of *The Essential Guide*. Note that distances (x) and (y) in the equation represent the distance from the wall to CR along the (x) and (y) directions, respectively.

$$F_{y,A} = \frac{R_{y,A}}{\sum R_y} \times F_y + \frac{M_{ta} \times (x) \times R_{y,A}}{\sum R_y (x)^2 + \sum R_x (y)^2}$$

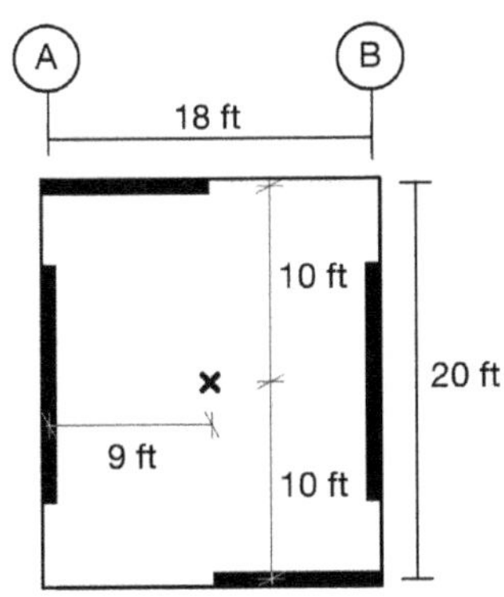

$$= \frac{1}{1+1} \times 75 + \frac{(\mp 67.5) \times (9) \times 1}{(1 \times 9^2 + 1 \times 9^2) + (1 \times 10^2 + 1 \times 10^2)}$$

$$= (35.8\ kip, 39.2 kip)$$

The maximum value that governs is therefore 39.2 kip.

Correct Answer is (B)

SOLUTION 1.39

Adding reinforcing steel to the inside or outside of the wall and applying new concrete effectively increases in-plane shear capacity, as it directly strengthens the wall against lateral forces as suggested by **Option B**.

Below is explains the incorrect options:

Option A: The thin surface treatment is primarily intended to improve out-of-plane flexural performance, not in-plane shear capacity.

Option C: The center coring technique is a developing method that primarily enhances vertical flexural capacity by creating reinforced cores (typically 4 in diameter placed at 3 ft to 4 ft intervals) acting as vertical beams. While it offers some benefit, it is not primarily intended for in-plane shear.

Option D: Filling in windows or doors may help with overall wall integrity but is highly dependent on architectural constraints and may not always be the best provided or architecturally feasible solution for this purpose.

Correct Answer is (B)

SOLUTION 1.40

Reference is made in this solution to SDPWS 2021, Table 4.3A. It is important to note that this table provides the nominal capacities for nailed wood shear walls; therefore, no other factors need to be applied to obtain these values (*).

The necessary calculation involves multiplying the value provided in this table by the length indicated on the sketch associated with this question as follows:

Wall	v_v (plf)	Length (ft)	V_n (kip)
1	505	12	6.06
2	950	12	11.4
3	670	9	6.03
			23.49

Correct Answer is (C)

(*) For additional information, consult *The Essential Guide* Section 6.3.4 where the method is described with more examples. It is also necessary to consider the notes below this table, such as using a different wood grade than Douglas Fir-Larch, or varying wood moisture content, and how these variables will influence the final shear capacity.

SOLUTION 1.41

According to ATC 20, page 98, the most frequent failure in traction elevators during seismic events is counterweights derailing from their guides. This can lead to severe damage or injury, especially if the counterweights impact the elevator cab or its occupants.

Correct Answer is (B)

SOLUTION 1.42

According to SDWPS 2021 Tables 4.2A to C, the factor is 0.5, and is used to account for the decrease in stiffness due to elevated moisture levels, as wood becomes more flexible and less stiff when its moisture content exceeds 19%.

Correct Answer is (A)

SOLUTION 1.43

For this question, refer to Section 6.2.4 of *The Essential Guide* (also found in CBC 2022, Section 1704.6.1). You should also consult the risk categories outlined in *The Essential Guide*, Section 6.2.2, pages 209–210 (also found in CBC 2022, Section 1604.5).

Based on this information, review the summary table provided for those structures.

Option	Risk category (if applicable)	Required observation
A	III	Yes
B	-	No
C	-	Yes
D	III	Yes

For this question, if you do not have a copy of *The Essential Guide*, you may refer to CBC 2022, Section 1704.6.1 for structural observations and Chapter 16 (Section 1604.5) for information on risk categories.

Correct Answer is (B)

SOLUTION 1.44

The method for determining the appropriate load case for the nailed diaphragm is explained in *The Essential Guide*, page 222.

This diaphragm has a staggered panel arrangement, where the continuous panel edges are perpendicular to the framing, and the applied load is also perpendicular to the framing direction. According to SDPWS Table 4.2B, this configuration corresponds to Load Case 3.

While this diaphragm includes the use of multiple rows of nails, based on the values in SDPWS 2021 Table 4.2B, and the details provided in the question, the nominal unit shear capacity is $2{,}115\ plf$.

To determine the allowable shear capacity, apply the shear reduction factor of 2.8 as described in *The Essential Guide*, Section 4.4.3 (*page 143*), or SDPWS 2021 Section 4.1.4. This gives the following allowable unit shear:

$$v_a = \frac{2{,}115}{2.8} = 755.4\ plf$$

You can find the information in this question in SDPWS 2021 Table 4.2B, and Section 4.1.4.

Correct Answer is (A)

SOLUTION 1.45

Pressure vessels are typically large enough to be classified as nonbuilding structures. However, this vessel's size and weight, which is less than 25% of the building seismic weight, can be treated as a nonstructural component. For further verification, refer to the diagram presented in Figure 5-6 of *The Essential Guide.*

Based on this information, the equation from Section 5.3.3 of *The Essential Guide* (Equation 13.3-1 from ASCE 7), along with Table 13.6-1 of ASCE 7, can be used to calculate the lateral seismic design force.

The following parameters are collected from the referenced table.

$$a_p = 2.5$$

$$R_p = 2.5$$

Moreover, z in this case is taken from the anchored base of the component as follows:

$$z/h = 1/3$$

$$F_p = \frac{0.4a_pS_{DS}W_p}{\left(\frac{R_p}{I_p}\right)}\left(1 + 2\frac{z}{h}\right)$$

$$= \frac{0.4 \times 2.5 \times 0.5 \times 4.5}{\left(\frac{2.5}{1}\right)}\left(1 + 2 \times \frac{1}{3}\right)$$

$$= 1.5\ kip$$

Check minimum and maximum values for the design force:

$$0.3S_{DS}I_pW_p \le F_p \le 1.6S_{DS}I_pW_p$$

$$0.3S_{DS}I_pW_p = 0.3 \times 0.5 \times 1 \times 4.5 = 0.68\ kip < 1.5\ kip \rightarrow ok$$

$$1.6S_{DS}I_pW_p = 1.6 \times 0.5 \times 1 \times 4.5 = 3.6\ kip > 1.5\ kip \rightarrow ok$$

$$M_{base} = 1.5\ kip \times \frac{9}{2}ft = 6.75\ kip.ft$$

You can find the information presented in this solution in the following sections/tables of ASCE 7:

- Section 15.3.1 to identify if this vessel is a nonstructural component or a nonbuilding structure.
- Equations 13.3-1, 2 & 3 for seismic design forces.
- Table 13.6-1 for nonstructural components coefficients.

Correct Answer is (C)

SOLUTION 1.46

Seismic coefficients for interior walls and nonstructural walls can be picked up from Table 13.6-1 of ASCE 7. These can be used to calculate the lateral seismic design force.

The following parameters are collected from the referenced table.

$$a_p = 1$$

$$R_p = 2.5$$

$$W_p = 12\ psf \times 9\ ft = 108\ plf$$

Where z/h in this case is taken from the anchored base of the component and equals to zero being at the first floor.

$$F_p = \frac{0.4a_p S_{DS} W_p}{\left(\frac{R_p}{I_p}\right)}\left(1 + 2\frac{z}{h}\right)$$

$$= \frac{0.4 \times 1 \times 0.9 \times 108}{\left(\frac{2.5}{1.5}\right)}(1 + 2 \times 0)$$

$$= 23.3\, plf$$

Check minimum and maximum values for the design force:

$$0.3S_{DS}I_pW_p \le F_p \le 1.6S_{DS}I_pW_p$$

$$0.3S_{DS}I_pW_p = 0.3 \times 0.9 \times 1.5 \times 108 = 43.7\, plf > 23.3\, plf$$

→ Use $F_p = 43.7\, plf$

You can find the information presented in this solution in the following sections/tables of ASCE 7:

- Equations 13.3-1, 2 & 3 for seismic design forces.
- Table 13.6-1 for nonstructural components coefficients.

Correct Answer is (B)

SOLUTION 1.47

Reference in this question is made to Table 5-2 of *The Essential Guide* where it states that the minimum operating weight W_p for a suspended ceiling should be $4\, psf$, which is greater than the operating weight provided in the question. Hence, a $W_p = 4\, psf$ governs.

Moreover, it is important to read the question carefully as the question only asks for the minimum seismic force, in which case, the minimum force equation provided in Section 5.3.3 of *The Essential Guide* will be used for this purpose. See below:

$$F_{p,min} \ge 0.3S_{DS}I_pW_p$$

$$\ge 0.3 \times 0.55 \times 1.0 \times 4\, psf$$

$$\ge 0.66\, psf$$

You can find the information presented in this solution in the following sections/tables of ASCE 7:

- Equations 13.3-3 for minimum seismic design forces.
- Section 13.5.6.1 for suspended ceilings.

Correct Answer is (C)

SOLUTION 1.48

Reference in this question is made to the following sections and tables from *The Essential Guide*:

- Section 5.4.2 that discusses base shear for nonbuilding structures.
- Section 6.2.2 page 210 which assigns power plants required as backup for risk categories IV (water storage facilities) to risk category IV, hence $I_e = 1.5$.
- Figure 5-6 which presents one of our custom diagrams that takes you through the steps of solving nonstructural components and nonbuilding structure questions.

Using the diagram presented in Figure 5-6, and since this chimney has $T > 0.06\, sec$:

$$V = C_s W$$

C_s can be calculated using Table 5-5 and 5-6 of *The Essential Guide.*

Since the question requests the minimum value and provided this is a chimney designed in accordance with a certain ACI requirement, with $S_1 > 0.6$ → Table 5-6 of the same section shall be used (*).

Also, R is taken from ASCE 7 Table 15.4-2 as $R = 2$.

$$C_{s,min} = \frac{0.5S_1}{\left(\frac{R}{I_e}\right)} = \frac{0.5 \times 0.62}{\left(\frac{2}{1.5}\right)} = 0.23$$

$$V = C_s W = 0.23 \times 11{,}450\ kip \approx 2{,}634\ kip$$

Information provided in this solution can be found in the following sections and codes:

- ASCE 7 Table 15.4-2.
- Risk categories can be found in CBC 2022 Section 1604.5.
- ASCE 7 Section 15.4.1, point 2, the exception section for chimney's designed in accordance with ACI 307.

Correct Answer is (A)

(*) Section 5.4.2 and Table 5-6 of *The Essential Guide* state that these values apply to the design of tanks, stacks, and chimneys that meet certain AWWA, API or ACI requirements. The list of requirements is detailed in ASCE 7 Section 15.4.1 point 2, page 147 under the EXCEPTIONS part of this section.

SOLUTION 1.49

This solution references Section 5.3.1 of *The Essential Guide.*

As stated in this section, and in alignment with ASCE 7 Section 13.1.4, there are several exemptions applicable when components are light in weight, and classified under SDC B. Other requirements include the location of the center of mass relative to the adjacent floor, as well as the importance factor.

In this context:

Option A: A non-load-bearing gypsum wall in SDC B is **exempt** under ASCE 7 13.1.4(3) architectural component.

Option B: A rooftop A/C unit in SDC B is **exempt** under ASCE 7 13.1.4(4) mechanical component.

Option C: A $300\ lb$ panel, with a center of mass $3.5\ ft$ above floor, SDC D, $I_p = 1.0$, and flexible connection, is **exempt** under ASCE 7 13.1.4(6)(a).

Option D: A $75\ lb$ panel, with a center of mass $4.5\ ft$ above floor, **which exceed the limit of $4\ ft$**, SDC D, $I_p = 1.0$, and flexible connection, is **not exempt** under ASCE 7 13.1.4(6)(a).

You can find the information presented in this solution in ASCE 7 Section 13.1.4 points 1 to 6.

Correct Answer is (D)

SOLUTION 1.50

This question uses *The Essential Guide,* Section 5.4.2 page 180, or Figure 5-6, which presents a simplified design process for nonbuilding structures. For further detail, see ASCE 7, Section 15.3.

According to the referenced design diagram, the monument qualifies as a nonbuilding structure with a weight greater than 25%, and a fundamental period less than $T < 0.06\ sec$, classifying it as a rigid element.

In this case, the supporting structure can be designed in accordance with either Chapter 12 or Section 15.5 of ASCE 7. Under both sections, ordinary steel moment frames are assigned a response modification factor $\boldsymbol{R} = \mathbf{3.5}$.

You can find the information presented in this solution in ASCE 7 Section 15.3 point 1.

Correct Answer is (A)

SOLUTION 1.51

Given a Site Class D and consulting with Tables 1-2 and 1-3 of *The Essential Guide* to determine values of F_a and F_v:

$$S_{DS} = \frac{2}{3} \times F_a \times S_s = \frac{2}{3} \times 1.15 \times 0.88 = 0.67$$

$$S_{D1} = \frac{2}{3} \times F_v \times S_1 = \frac{2}{3} \times 1.7 \times 0.65 = 0.74$$

$$T_s = \frac{S_{D1}}{S_{DS}} = \frac{0.74}{0.67} = 1.1 < T\ (= 1.6)\ (*)$$

Moreover, use Table 15.4-1 of the ASCE 7 to collect the response modification factor of the mentioned SFRS as $R = 1$ using the unlimited height for an SDC D.

Using Table 5-4 & 5 of *The Essential Guide*:

$$V = C_s W$$

$$C_s = \frac{S_{D1}}{T\left(\frac{R}{I_e}\right)} = \frac{0.74}{1.6\left(\frac{1}{1}\right)} = 0.46$$

Check $C_{s,min}$ using the following equation because $S_1 > 0.6$:

$$C_{s,min} = \frac{0.8S_1}{\left(\frac{R}{I_e}\right)} = \frac{0.8 \times 0.65}{\left(\frac{1}{1}\right)} = 0.52$$

→Use $C_s = 0.52$

$$V = C_s W = 0.52 \times 120 = 62.4\ kip$$

The information mentioned above can be found in ASCE 7 Sections:

- S_{DS} & S_{D1}: Section 11.4.5.
- I_e: Table 1.5-2.
- F_a & F_v: ASCE 7 Table 11.4-1 & 2.
- R: Table 15.4-2.
- $C_{s,min}$: Eq. 15.4-2 when $S_1 \geq 0.6g$.

Correct Answer is (B)

(*) If $T > 1.5T_s$ then C_s should be multiplied by 1.5. See ASCE 11.4.8 exception 2 when $S_1 \geq 0.2$ for SDC D. Or the *Essential Guide* Notes Section, Note#1.

SOLUTION 1.52

The Essential Guide provides detailed information on convective and impulsive loads in page 185. See below:

I. The convective response can govern the design of the tank roof.

The sloshing (convective) motion of the liquid exerts significant hydrodynamic pressure under the tank roof. In large or partially full tanks, this sloshing can become intense which may threaten the structural integrity of the roof. → **True.**

II. The total lateral load on a tank is typically divided into impulsive and convective components.

The total lateral load is divided into impulsive and convective components to account for the different dynamic behaviors of the liquid. The impulsive part represents the liquid that moves with the tank wall, while the convective part accounts for the sloshing motion. This separation allows for more accurate seismic design. → **True.**

III. The convective mode has a shorter natural period compared to the impulsive mode.

The sloshing of the liquid (convective) responds more slowly than the impulsive mode, which moves with the tank shell. Because of this slower, wave-like motion, the convective mode has a much longer natural period compared to the impulsive mode. → **Untrue.**

IV. Impulsive loads tend to be more significant in smaller tanks, whereas convective become more dominant in larger tanks.

In smaller tanks, the majority of the liquid moves with the tank wall, making the impulsive load the dominant component in seismic design. However, in larger tanks, more of the liquid exhibits sloshing behavior, increasing the significance of the convective load. → **True.**

You can find information on convective and impulsive loads in ASCE 7 Section 15.7.6.

Correct Answer is (D)

SOLUTION 1.53

A detailed discussion on special moment frames and RBS is provided in Section 6.6 of *The Essential Guide.*

An RBS is a portion of the beam flange that is intentionally cut away/reduced near the beam-column connection to create a controlled weak zone. This encourages plastic hinging to occur in the beam, away from the connection, enhancing ductility and protecting the column and welds during earthquakes – **this makes Options A, C and D correct**. See figure below for better understanding.

The protected zone is the area expected to undergo inelastic deformation during an earthquake. These zones are deliberately left unaltered and clear of any attachments, brackets, holes or supports – hence the name. Their purpose is to enable ductile behavior of the framing structure. Consequently, the RBS is part of the protected zone – **this makes Option B incorrect.**

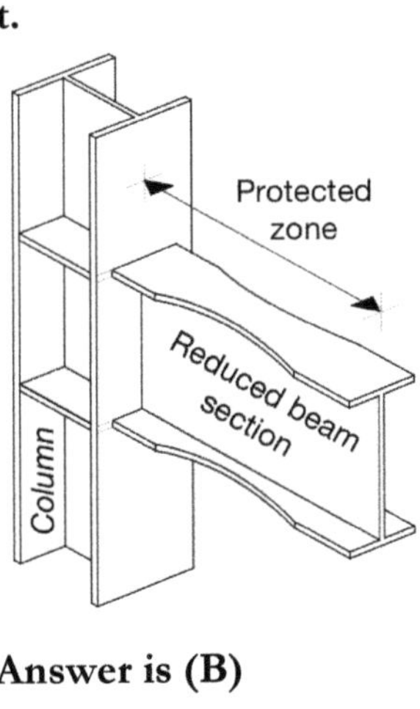

Correct Answer is (B)

SOLUTION 1.54

Section 6.5.3 of *The Essential Guide* covers seismic detailing of concrete columns and beams in special moment frames.

Transverse reinforcement in reinforced concrete columns – such as ties or spirals – serves several critical functions in seismic design. It confines the core concrete enhancing the ductility and compressive strength of the column under earthquake loading **(Option A)**. It also provides lateral support to the longitudinal reinforcement bars preventing them from buckling during column swaying and deformation **(Option B)**. Additionally, transverse reinforcement confines the longitudinal bars at lap splice locations improving performance under cyclic loading **(Option C)**.

Although the above may indirectly improve moment capacity – or, in other words preserves it – it does not significantly improve it, which is primarily governed by longitudinal reinforcement along with the column cross section.

Columns subjected to seismic loading fail in shear if transverse reinforcement is inadequate as they enhance shear strength and improve the overall ductility.

This makes **Option D** the correct selection being the wrong option from the above.

Correct Answer is (D)

SOLUTION 1.55

You can find this information in the Notes section of *The Essential Guide.*

Critical damping is the level of damping that prevents a building from oscillating. A critically damped system returns to its original position without oscillations.

This principle is the basis behind seismic design. In practice, most buildings are designed with a fraction of critical damping (e.g., 5%), which helps reduce vibration magnitude during seismic events.

Correct Answer is (C)

BONUS QUESTION SOLUTION

The CBC 2022 permits one story only in light wood frame construction in SDC D & E.

This along with many other requirements are summarized in Table 6-5 of *The Essential Guide.*

> You can find the information presented in this solution in CBC 2022 Table 2308.2.1.

Correct Answer is (A)

EXAM TWO

Answer Sheet

No.	Your Answer	Fill our Answer	Notes
2.1			
2.2			
2.3			
2.4			
2.5			
2.6			
2.7			
2.8			
2.9			
2.10			
2.11			
2.12			
2.13			
2.14			
2.15			
2.16			
2.17			
2.18			
2.19			
2.20			
2.21			
2.22			
2.23			
2.24			
2.25			
2.26			
2.27			
2.28			
2.29			
2.30			
2.31			
2.32			
2.33			
2.34			
2.35			
2.36			
2.37			
2.38			
2.39			
2.40			
2.41			
2.42			
2.43			
2.44			
2.45			
2.46			
2.47			
2.48			
2.49			
2.50			
2.51			
2.52			
2.53			
2.54			
2.55			

PROBLEM 2.1

What is the site classification for a location that has a shear wave velocity of 1,250 ft/s for the first 100 ft of depth?

(A) Class C – Dense soil or soft rock

(B) Class D – Stiff soil

(C) Class E – Soft clay soil

(D) Class F – Vulnerable to failure

PROBLEM 2.2

After a significant earthquake, several slopes in a coastal region failed. Based on the following scenarios, which slope is most likely to have failed during the earthquake?

(A) A moderately steep slope near a coastal cliff, composed of loose, silty sand with a high-water table.

(B) A steep slope in a mountainous area with weathered limestone and active cracking, but no recent erosion.

(C) A gentle slope with dry clay, recently disturbed by construction near a drainage ditch with a low water table.

(D) A steep slope in a semi-arid region, composed of well-graded gravel, with low soil moisture content.

PROBLEM 2.3

A site is being evaluated for seismic design where its first 150 ft consists of medium stiff clay with an average undrained shear strength $s_u = 900\,psf$. What should be the site classification for this location?

(A) Site Class C

(B) Site Class D

(C) Site Class E

(D) Site Class F

PROBLEM 2.4

A detention facility that will be located in a Site Class C has the following data collected from the ASCE online hazard tool:

- $S_s = 1.25$
- $S_1 = 0.75$

Based on the above information, what is the best Seismic Design Category (SDC) for this building?

(A) SDC C

(B) SDC D

(C) SDC E

(D) SDC F

PROBLEM 2.5

An acute care hospital under the OSHPD (Office of Statewide Health Planning and Development) in San Francisco with no emergency surgery facility, is planned to be 100 ft tall. The following information has been gathered for the proposed building location:

- $S_s = 1.0$
- $S_1 = 0.2$
- Site Class E

Based on the above information, what is the best Seismic Force Resisting System (SFRS) for a Moment Resisting Frame System?

(A) Steel special truss moment frames.

(B) Steel intermediate moment frames.

(C) Steel ordinary moment frames.

(D) Steel and concrete composite partially restrained moment frames.

PROBLEM 2.6

Which one of the below systems is best fit to build a $45\,ft$ tall single story height storage warehouse facility in an SDC D in a building frames system?

(A) Steel eccentrically braced frames.

(B) Intermediate precast shear walls.

(C) Ordinary precast shear walls.

(D) Special reinforced masonry shear walls.

PROBLEM 2.7

The below building consists of nine floors each floor is $14\,ft$ tall, and the building is assigned to SDC C. The SFRS is made of intermediate reinforced masonry shear walls.

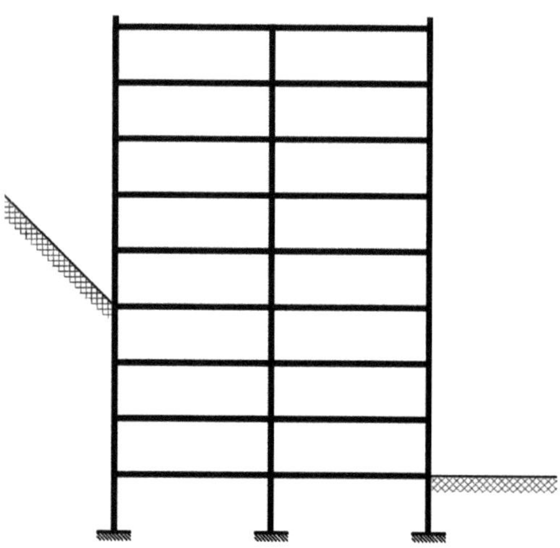

As shown in the above sketch, the first floor is below grade (i.e., basement), and to the left of the building there is a steep slope supported by the building's retaining walls.

What is the approximate fundamental period for this building?

(A) $0.48\,sec$

(B) $0.62\,sec$

(C) $0.69\,sec$

(D) $0.75\,sec$

PROBLEM 2.8

In order to create more space for a single-degree-of-freedom, single-story building, a few shear walls were reduced without impacting the effective weight or the required lateral strength of the structure. This has reduced the stiffness of the building in the direction under study by half.

How will this affect the fundamental period of the modified structure? And, what impact will it have on the seismic forces acting on the building?

(A) $T_{new} = \sqrt{2} \times T_{current}$

The lateral force reduces for the modified structure.

(B) $T_{new} = \sqrt{2} \times T_{current}$

The lateral force increases for the modified structure.

(C) $T_{new} = \frac{1}{\sqrt{2}} \times T_{current}$

The lateral force reduces for the modified structure.

(D) $T_{new} = \frac{1}{\sqrt{2}} \times T_{current}$

The lateral force increases for the modified structure.

PROBLEM 2.9

To enhance the ductility of a steel ordinary moment resisting frame in a seismic force-resisting system, which of the following strategies should be considered?

(A) Combine it with a more ductile system in the direction of interest.

(B) Must satisfy a certain strong-column/weak-beam ratio.

(C) Use lower structural steel grade in the moment frame.

(D) Use higher structural steel grade in the moment frame.

PROBLEM 2.10

A bridge will be connecting Building 1 (left) and building 2 (right) at the roof level of building 1. The two buildings are assigned to SDC C.

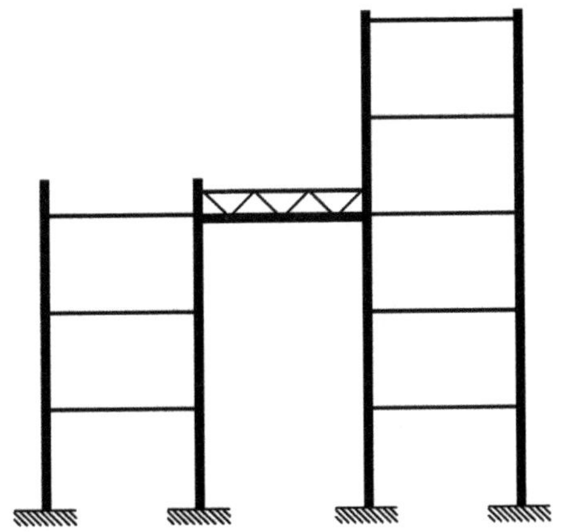

Building 1 (left):

- Building with public assemblies with occupant load of 250 per assembly, and a total occupant load of 2,000.
- The building is made of steel and concrete composite ordinary moment frame.

Building 2 (right):

- A mall with an occupant load of 5,500.
- The building is made of intermediate reinforced concrete moment frame.

Elastic deflections for the two buildings are as follows:

Floor No	Building 1	Building 2
1	1.9 *in*	1.7 *in*
2	2.8 *in*	2.7 *in*
3	3.5 *in*	3.7 *in*
4	–	4.6 *in*
5	–	5.5 *in*

What distance should the bridge support be designed for?

(A) 7.2 *in*

(B) 12.3 *in*

(C) 15.9 *in*

(D) 38.0 *in*

PROBLEM 2.11

The following sketch is for a rigid rectangular diaphragm which belongs to an SDC C. The Center of Mass (CM) is located at the center of the diaphragm, while the Center of Rigidity (CR) is 0.5 ft away from the center of mass as shown below.

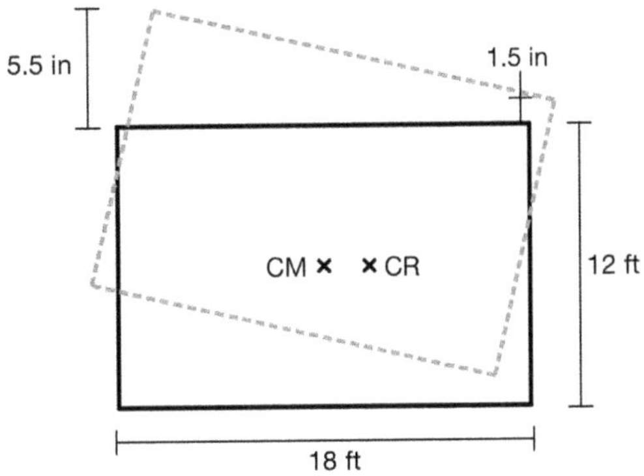

The load applied in the short direction of this slab is 75 kip.

What is the total torsional moment for this diaphragm?

(A) 37.5 $kip.ft$

(B) 131.25 $kip.ft$

(C) 152.25 $kip.ft$

(D) 178.5 $kip.ft$

PROBLEM 2.12

For an acute care hospital regulated by OSHPD (Office of Statewide Health Planning and Development), assigned to Seismic Design Category D and exhibiting a Type 1.b extreme soft story irregularity, what is the most appropriate method of seismic analysis for this structure?

(A) This irregularity is prohibited for the mentioned SDC.

(B) More information on height and fundamental period is required to make this determination.

(C) Equivalent lateral force procedure.

(D) Nonlinear response history analysis.

PROBLEM 2.13

The diaphragm design seismic forces F_{px} specified in ASCE 7 are intended to account for which of the following?

(A) First mode of vibration along with localized diaphragm demands.

(B) Combination between first, second and third modes.

(C) Wind load in combination with first mode of vibration.

(D) Higher modes of vibration effects and localized diaphragm demands.

PROBLEM 2.14

What is the most common type of seismic failure observed in older houses and dwellings?

(A) Bowing of underground concrete walls.

(B) Foundations cracking.

(C) Failing of spandrel beams.

(D) House coming off the foundation.

PROBLEM 2.15

This sketch is for a pilaster supported wall and each pilaster is anchored from its top.

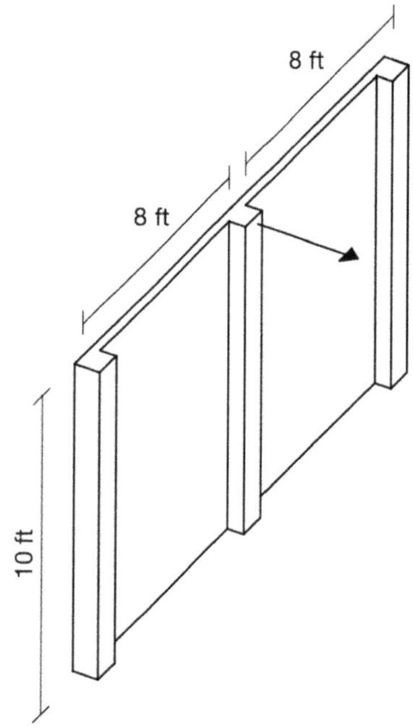

Based on the above description, what is the walls tributary area that should be considered to calculate seismic loads for the middle top anchor as shown in the sketch?

(A) $20\ ft^2$

(B) $32\ ft^2$

(C) $48\ ft^2$

(D) $64\ ft^2$

PROBLEM 2.16

The below is a tilt-up concrete wall that is part of a power generation station. The wall thickness is $9\ in$, and the unit weight of concrete is $150\ pcf$.

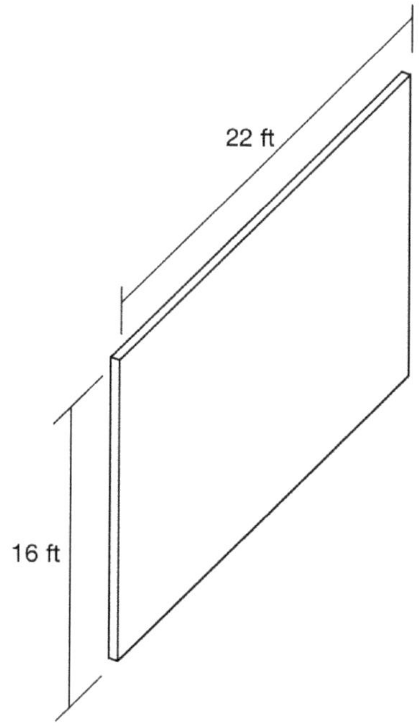

The Site Class is D, with seismic parameters $S_s = 1.1$ and $S_1 = 0.4$.

What is the total seismic out-of-plane force generated by this wall?

(A) 17,989 lb

(B) 15,444 lb

(C) 5,543 lb

(D) 3,960 lb

PROBLEM 2.17

The below is a single story 14 ft tall building with a concrete slab 12 in thick and five reinforced concrete shear walls as shown, each one is 10 ft wide and 9 in thick.

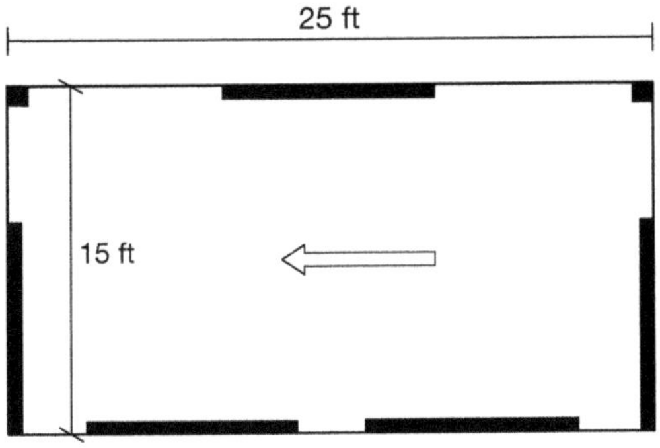

Assuming that the $C_s = 0.85$ and the reinforced concrete used has a unit weight of 150 pcf, what is the base shear in the direction shown?

(A) 61,200 lb

(B) 67,894 lb

(C) 72,000 lb

(D) 81,281 lb

PROBLEM 2.18

Which of the following buildings can use a redundancy factor of 1.0, given that no horizontal irregularities are present and the structure is located in SDC D?

(A) A building with one or more bays of the SFRS is resisting its lateral loads.

(B) A two-story building with cantilever columns where the loss of resistance at any column base does not result in > 33% reduction in story lateral strength.

(C) A story resisting 25% of base shear with braced frames, where the removal of one brace does not lead to > 33% reduction in story lateral strength.

(D) A story resisting 25% of base shear with moment frames, where the loss of moment capacity in a beam-column connection does not result in more than a 33% reduction in story strength.

PROBLEM 2.19

Which of the following cross-sections best illustrates the shear transfer mechanism between a precast concrete diaphragm and a masonry shear wall?

(A) Option A

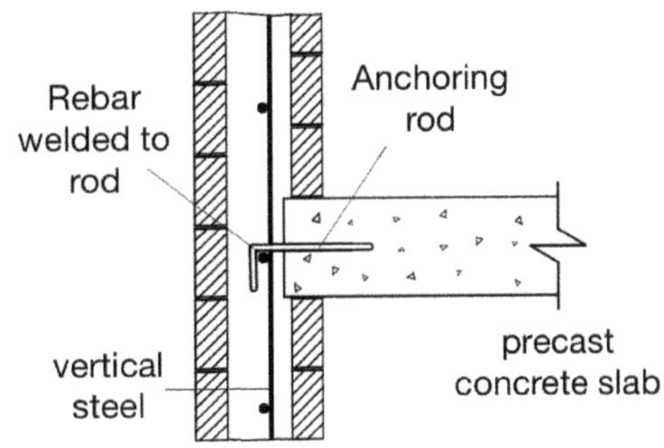

(B) Option B

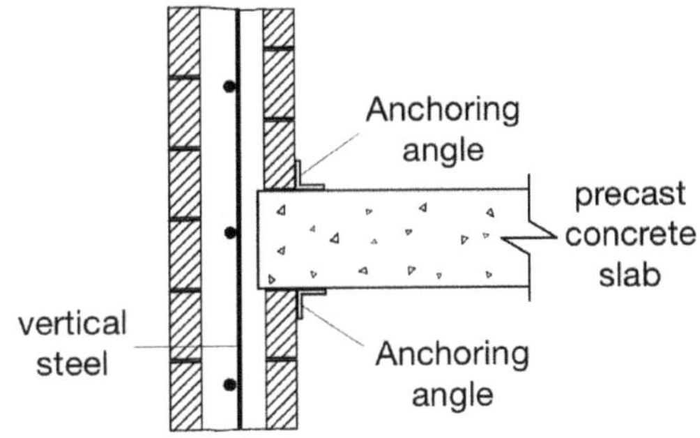

(C) Option C

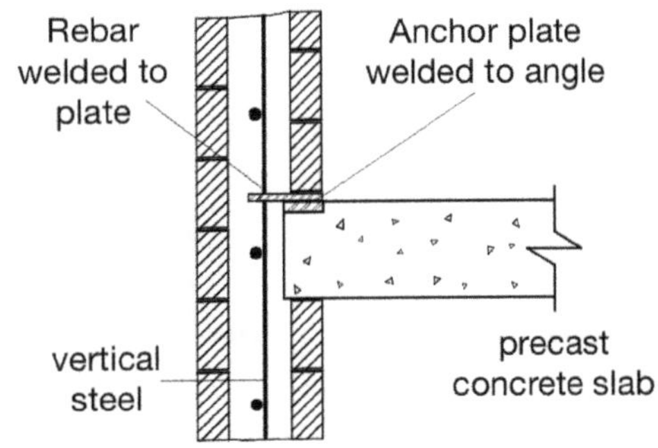

(D) Option D

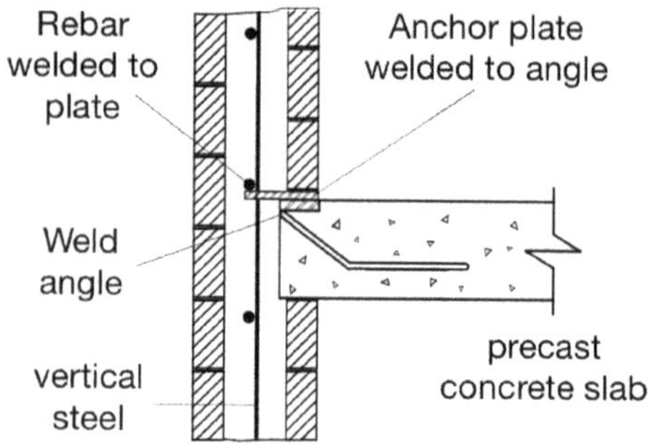

PROBLEM 2.20

A three-story building, with each story measuring 11 ft in height. The first story uses cantilevered exposed timber frame system, while the second and third stories use a building frame system as the seismic force-resisting system. The upper two stories consist of light wood frame walls sheathed with wood structural panels rated for shear resistance, designed for Seismic Design Category D.

Based on this configuration, what response modification factor should be used for analyzing the second and third floors?

(A) 1.5

(B) 4.5

(C) 7

(D) 9

PROBLEM 2.21

For the rigid diaphragm shown below, assume the origin of the x- and y-axes is located at the bottom-left corner.

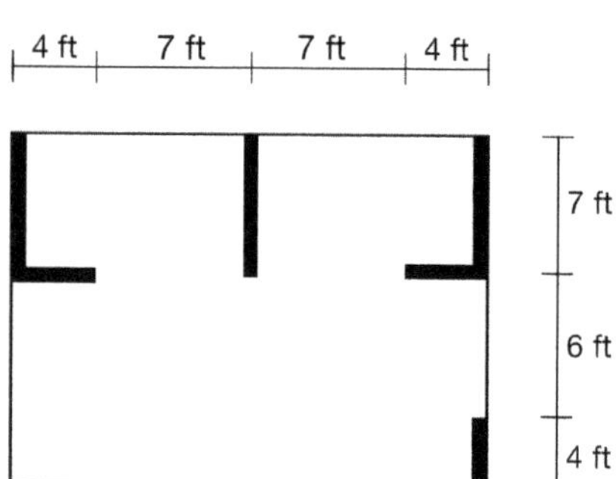

All walls are 12 in thick and 9 ft tall concrete walls, and they all can be assumed as cantilever walls.

What are the x and y coordinates of the center of rigidity (*)?

(A) $(12.1, 7.5)$

(B) $(11.8, 6.7)$

(C) $(11.2, 5.7)$

(D) $(11.2, 6.7)$

(*) Assume that the orthogonally connected shear walls are not structurally connected, as considering their combined stiffness is beyond the scope of this book.

PROBLEM 2.22

For the rigid diaphragm shown below, assume the origin of the x- and y-axes is located at the bottom-left corner.

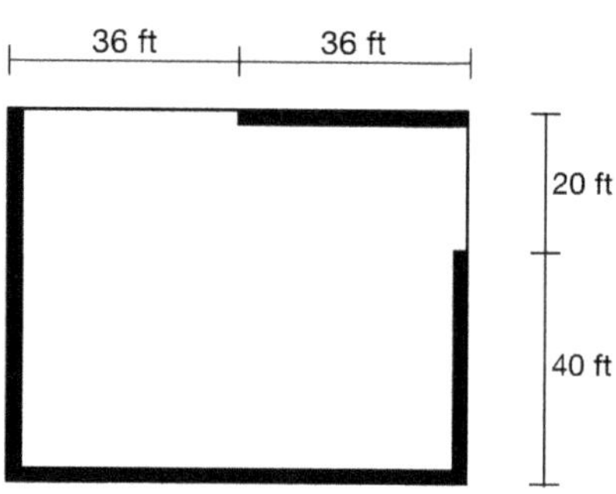

All walls are 12 in thick and 9 ft tall concrete walls, and they all can be assumed as cantilever walls.

What are the x and y coordinates of the center of rigidity (*)?

(A) $(18.9, 22)$

(B) $(28.8, 20)$

(C) $(32.4, 25)$

(D) $(22, 22)$

(*) Assume that the orthogonally connected shear walls are not structurally connected, as considering their combined stiffness is beyond the scope of this book.

PROBLEM 2.23

The building shown below shows its base shear distributed vertically. All floors have the same effective weight, with $S_{DS} = 0.65$. Disregarding maximum and minimum diaphragm design forces, what is the diaphragm design force at the roof level?

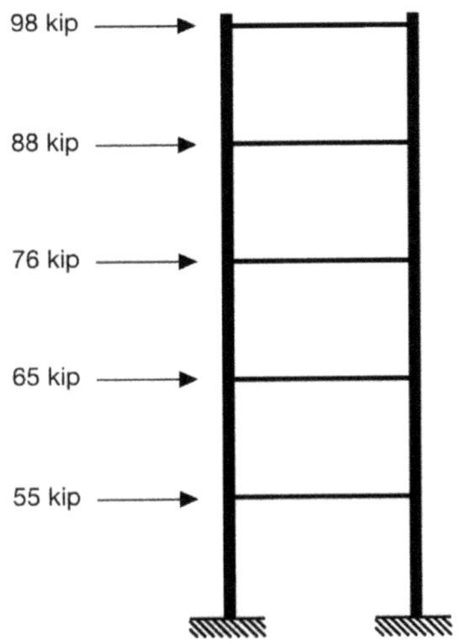

(A) $98\ kip$

(B) $128\ kip$

(C) $148\ kip$

(D) More information is needed.

PROBLEM 2.24

A 10-story building uses concrete moment-resisting frame as its SFRS. Each story is $10\ ft$ tall, and the effective seismic weight per story is equivalent. Use the following seismic parameters for this building:

- $S_{DS} = 0.65$
- $S_{D1} = 0.85$
- $T_{calculated} = 2.5\ sec$

Using this information, what is the vertical distribution factor C_{vx} for the 2nd floor:

(A) 0.01

(B) 0.02

(C) 0.03

(D) 0.04

PROBLEM 2.25

What is the minimum out-of-plane force that must be considered for anchoring an interior bearing wall using the simplified method?

(A) 10% weight of wall

(B) 10% weight of wall $\times I_e$

(C) 5% weight of wall $\times S_{DS}$

(D) 20% weight of wall $\times I_e$

PROBLEM 2.26

The building shown below is single-story with two types of reinforced concrete columns. The exterior columns measure $12\ in \times 24\ in$, while the middle column is $14\ in \times 14\ in$. The building is $12\ ft$ tall, and the plan dimensions are as indicated below. The concrete modulus of elasticity is $4.2 \times 10^3\ ksi$.

Columns can be modeled as fixed at the top and the bottom. The slab is $12\ in$ thick with a density of $150\ pcf$ and is the only one contributing to the seismic weight of the building.

Based on this information, determine the fundamental period of the structure in the direction indicated by the dotted arrow.

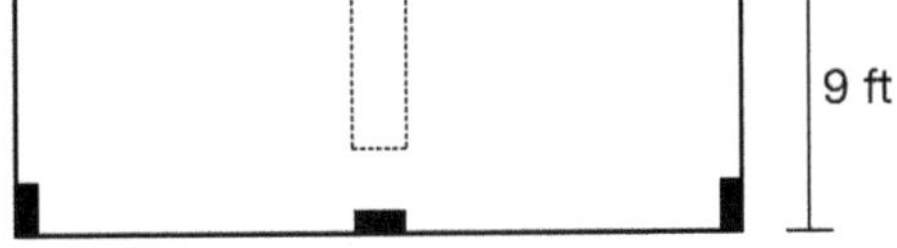

(A) $0.06\ sec$

(B) $0.12\ sec$

(C) $0.24\ sec$

(D) $0.36\ sec$

PROBLEM 2.27

The below diaphragm joists are connected perpendicular to the shear wall. Which of these connections does not provides an effective load path for shear force transfer?

(A) Option A

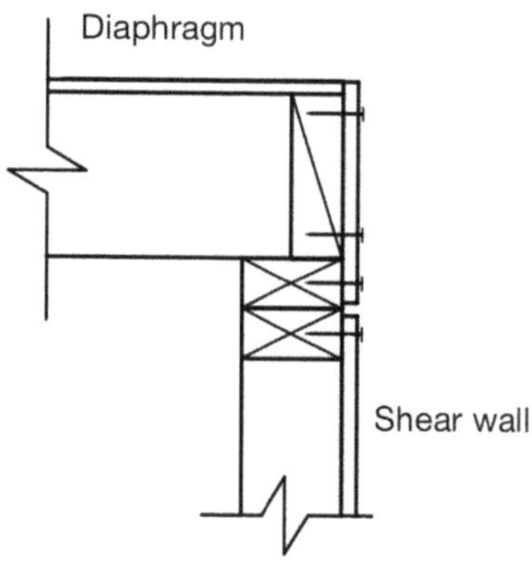

(B) Option B

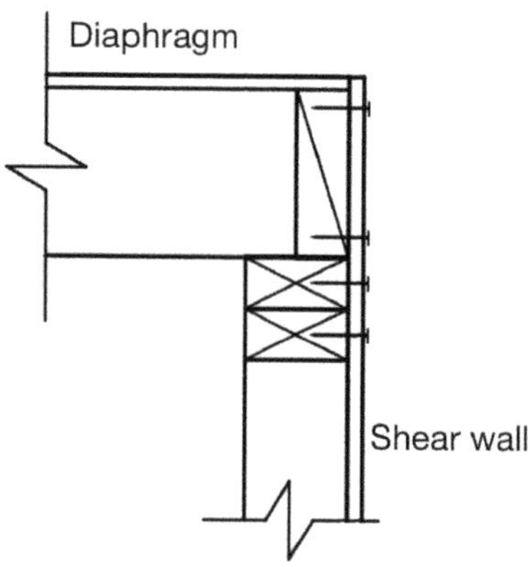

(C) Option C

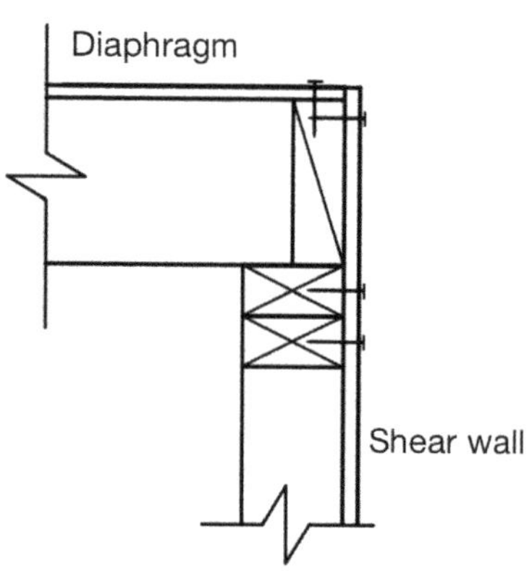

(D) Option D

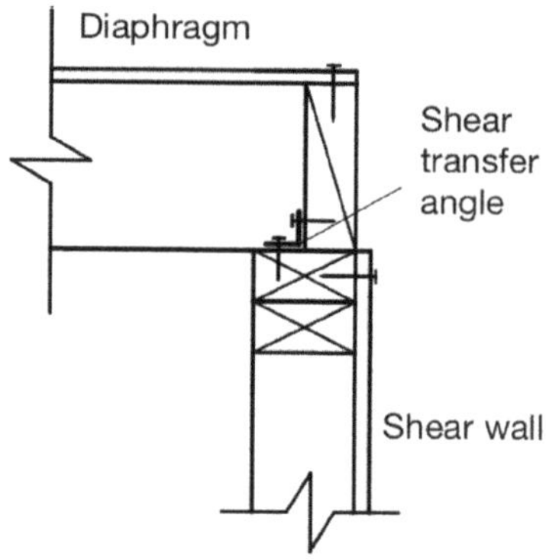

PROBLEM 2.28

A structural element that is part of a special shear wall SFRS system supports a building with an extreme torsional irregularity. The building is assigned to an SDC D and is subjected to the following loads:

- Dead load (D)
- Live load (L)
- Seismic horizontal load effect (Q_E)
- $S_{DS} = 0.5$
- $S_{av} = 0.3$
- $S_{D1} = 0.5$

Based on this information, which of the following equations best represent the ASD load combination that includes seismic effects?

(A) $0.91D + 0.55Q_E$

(B) $1.05D + 0.68Q_E + 0.75L$

(C) $1.05D + 0.88Q_E$

(D) $1.05D + 0.48Q_E + 0.75L$

PROBLEM 2.29

What is the vertical seismic effect on a structural element in an SDC B building with $S_{DS} = 0.5$, $S_{D1} = 0.7$, $S_{av} = 0.4$, and a dead load of 50 kip?

(A) 5 kip

(B) 6 kip

(C) 7 kip

(D) 0

PROBLEM 2.30

What is the tension in the chord at the point indicated on the right side below?

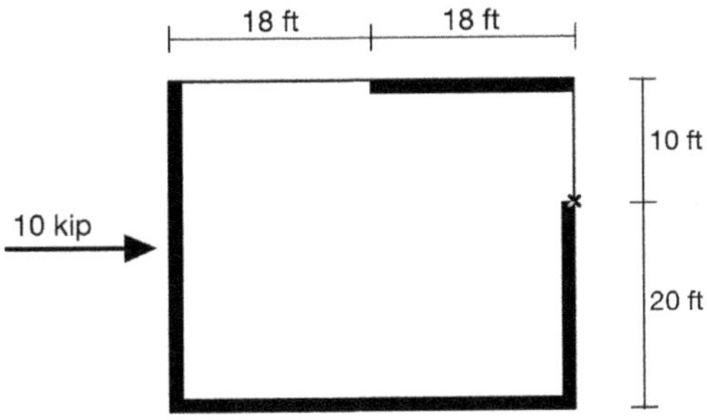

(A) 1.4 *kip*

(B) 1.9 *kip*

(C) 2.5 *kip*

(D) 3.2 *kip*

PROBLEM 2.31

The sketch below shows a wood roofing diaphragm connected to a masonry shear wall, with the main diaphragm framing perpendicular to the wall.

In this arrangement, which of the following acts as the chord of the diaphragm?

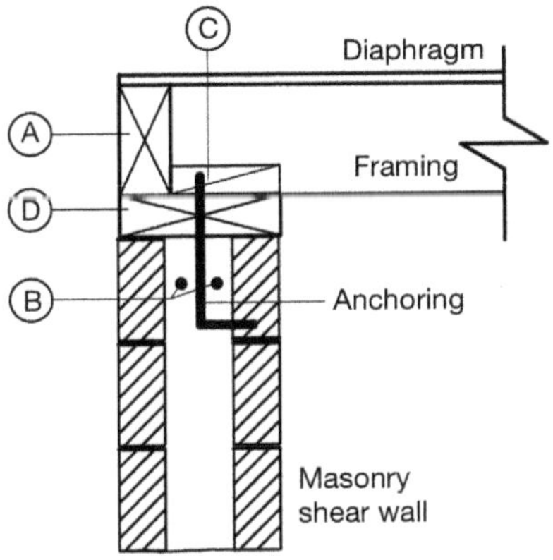

(A) The blocking of the main frame.

(B) Horizontal reinforcement inside wall, near, or at diaphragm level.

(C) Added blocking to wall top plate, also serves to connect the wall top plate to the diaphragm framing.

(D) Added top framing for the wall.

PROBLEM 2.32

What is the maximum height of a concrete moment-resisting frame building in SDC D, risk category II, no structural irregularity, with $S_{DS} = 0.5$ and $S_{D1} = 0.3$, that would still allow the use of the Equivalent Lateral Force (ELF) procedure for seismic analysis?

(A) $160\ ft$

(B) $175\ ft$

(C) $225\ ft$

(D) $323\ ft$

PROBLEM 2.33

What is the tension force in the right-side brace, given that the braced frame on the left is twice as rigid as the braced frame on the right, assuming all joints are pinned connections that carry no moment?

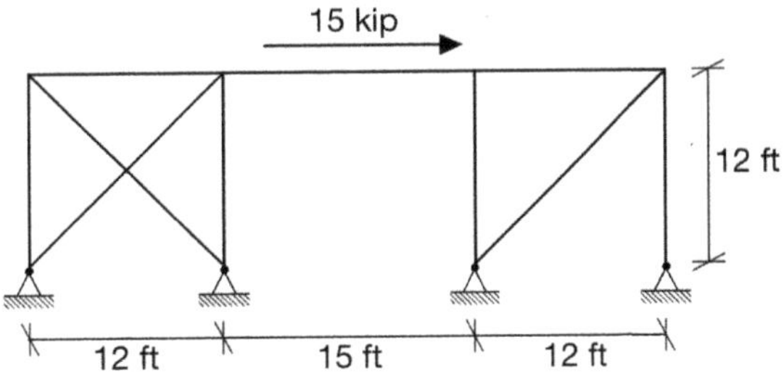

(A) 21.2 *kip*

(B) 14.1 *kip*

(C) 7.1 *kip*

(D) 5.9 *kip*

PROBLEM 2.34

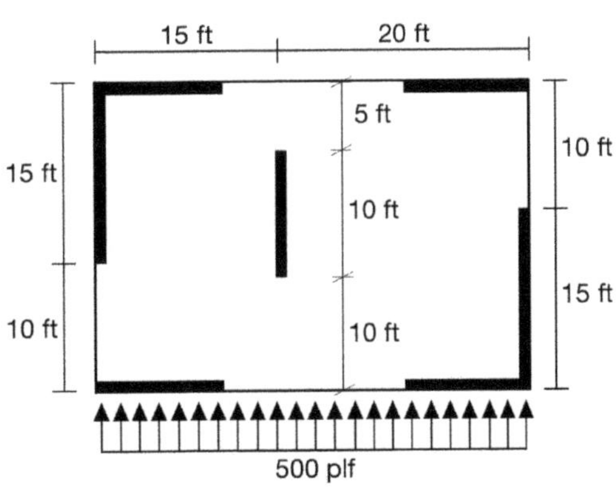

What is the maximum drag force for the single-story flexible diaphragm in the load direction?

(A) 1.5 *kip*

(B) 1.75 *kip*

(C) 2.0 *kip*

(D) 3.5 *kip*

PROBLEM 2.35

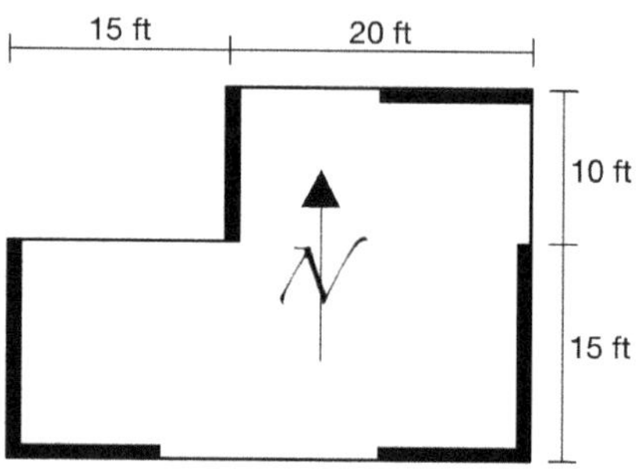

In the above shown flexible single-story diaphragm, what is the maximum drag force for a lateral diaphragm force applied in the North direction of a total of 50 *kip*?

(A) 7 *kip*

(B) 15 *kip*

(C) 18 *kip*

(D) 21 *kip*

PROBLEM 2.36

For the below single story, flexible diaphragm supported by shear walls, what is the load share at the wall at axis B for a total diaphragm force of 50 *kip* acting in the North direction?

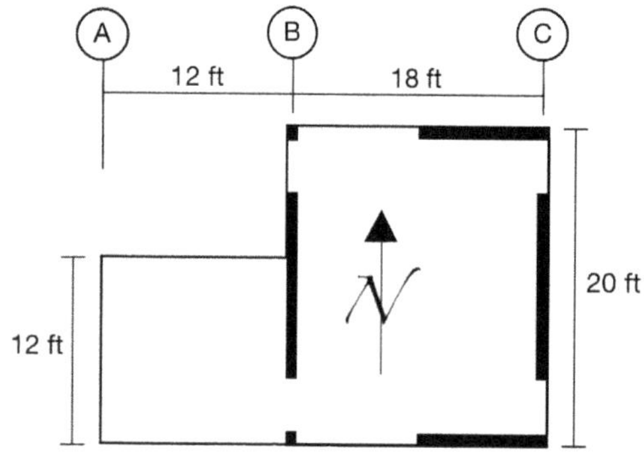

(A) 25.2 *kip*

(B) 32.2 *kip*

(C) 38.9 *kip*

(D) 41.1 *kip*

PROBLEM 2.37

For the below single story, flexible diaphragm supported by shear walls, what is the load share at the wall at axis C for a total diaphragm force of 50 *kip* acting in the North direction ignoring the lateral stiffness of columns at line B?

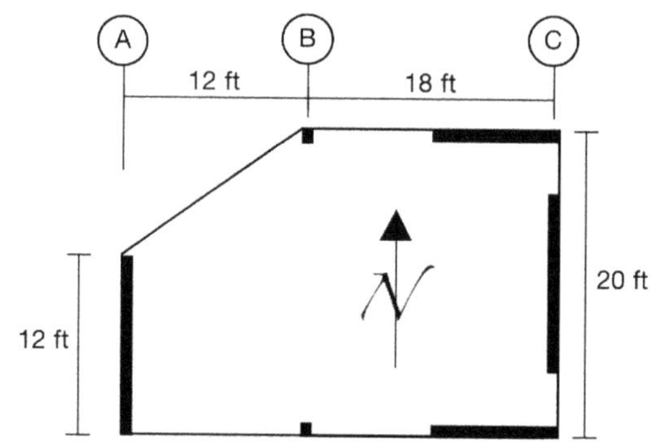

(A) 26.7 *kip*

(B) 28. 2*kip*

(C) 29.7 *kip*

(D) 33.1 *kip*

PROBLEM 2.38

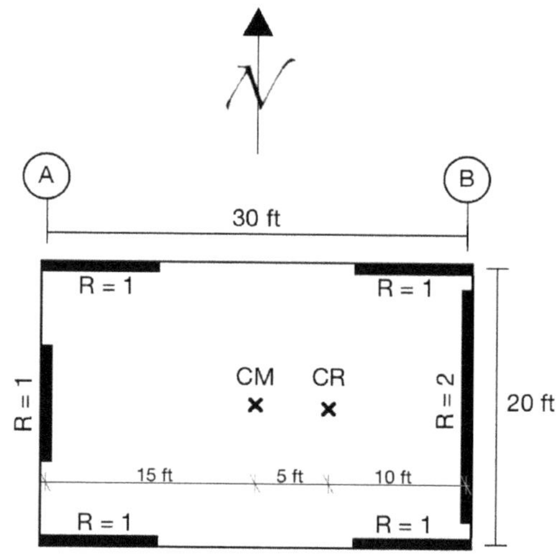

The rigid diaphragm shown here includes four shear walls with relative rigidities for walls shown on the sketch.

The Center of Mass (CM) is located at the centroid of the diaphragm and the Center of Rigidity (CR) has been calculated as indicated on the diaphragm.

Based on this, what is the load transferred to the wall at axis A for a lateral load applied in the North direction of 100 *kip*?

(A) 33.33 *kip*

(B) 39.33 *kip*

(C) 46.33 *kip*

(D) 52.33 *kip*

PROBLEM 2.39

In a properly designed eccentric braced frame (EBF) system, which of the following is the most likely mode of inelastic behavior or failure under lateral loading?

(A) Buckling of compression brace members.

(B) Yielding of compression brace members.

(C) Tension yielding of brace members

(D) Flexural and shear yielding of link beam members.

PROBLEM 2.40

The walls below are made from Douglas-Fir-Larch with structural I sheathing panels provided.

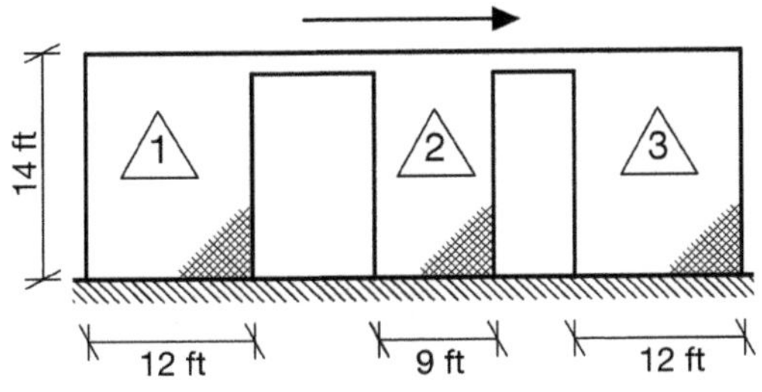

Wall	Nominal panel thickness (*in*)	Staple length and gauge (*in*)	Fastener spacing at panel edge (*in*)
1	3/8	1 – ½ 16	6 *in*
2	7/16	2 16	4 *in*
3	15/32	1 – ½ 16	3 *in*

What is the allowable shear capacity for the three shear walls combined in the direction shown?

(A) 8.5 *kip*

(B) 15.0 *kip*

(C) 17.5 *kip*

(D) 21.8 *kip*

PROBLEM 2.41

What is a key recommendation when inspecting a key hospital after an earthquake that you found as unsafe?

(A) Use a yellow placard and keep using the hospital but with care.

(B) Evaluate only the primary structural system if no obvious damage is present.

(C) Consult with appropriate facility staff before making recommendations that could limit the use of the facility.

(D) Shut down the hospital until a detailed engineering analysis is complete and move everyone to a safer place.

PROBLEM 2.42

Which of the following statements is NOT true regarding masonry nonparticipating walls?

(A) Nonparticipating walls are named as such because they are not part of the seismic resisting system.

(B) Nonparticipating walls should be isolated from the seismic resisting system so that the building lateral load does not impact them.

(C) Nonparticipating walls do not contribute to the lateral resisting system, but they may also resist vertical loads.

(D) Nonparticipating walls are normally unreinforced walls.

PROBLEM 2.43

Following an earthquake of magnitude 6.0 or greater, which of the following facilities is least likely to require prioritization for a detailed structural assessment?

(A) Community hospital main entrance building.

(B) 911 dispatch center parking garage.

(C) Emergency operations Center.

(D) A communication center.

PROBLEM 2.44

The sketch below shows a blocked wood diaphragm, with vertical blocking strips as illustrated. The diaphragm uses structural I sheathing, installed using 1 ½ in 16 gauge staples with 1 in penetration. The staples are spaced at 6 in on center throughout. The sheathing thickness is $3/8$ in, and the width of the nailed face of the framing members adjoining the panels is 3 in.

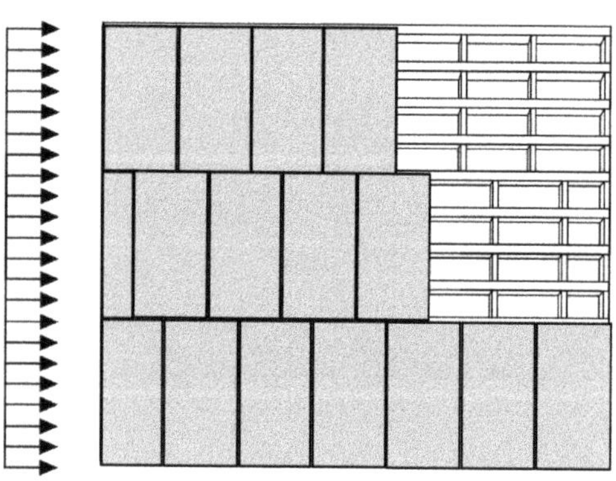

Based on this information, what is the allowable shear capacity in plf for the load acting in the direction indicated in the sketch if this diaphragm was made of Southern Pine wood?

(A) 175 plf

(B) 200 plf

(C) 160 plf

(D) 180 plf

PROBLEM 2.45

A ¾ in thick tempered glass cantilever parapet is planned to be installed on the second floor of a four equal height stories of a main hospital facility. The parapet is 4 ft high, weighs 9.5 psf, and is anchored only at its base. It is being evaluated for anchor design under seismic loading. The seismic design parameters are as follows:

- $S_{DS} = 0.6$
- $S_{D1} = 0.4$

What is the approximate seismic lateral design force generated at its base to be used for anchor assessment?

(A) 5.2 plf

(B) 8.2 plf

(C) 10.3 plf

(D) 20.5 plf

PROBLEM 2.46

A medium-sized rooftop Air Handling Unit weighing 7,000 lb is installed on a building rooftop. The unit is mounted on spring-type vibration isolators that provide a ½ in nominal air gap to allow for proper vibration isolation during normal operation.

The Importance Factor, $I_p = 1.0$, and the site seismic parameters are $S_{DS} = 0.3$ and $S_{D1} = 0.2$.

Assuming the unit is adequately anchored, and the isolators have horizontal restraints, what should be the seismic design force acting on the equipment?

(A) 0.63 kip

(B) 1.05 kip

(C) 1.26 kip

(D) 2.1 kip

PROBLEM 2.47

What percentage of equipment weight that are not fastened to access floors should be included in the operating weight of those floors for seismic force calculations?

(A) 0%

(B) 10%

(C) 15%

(D) 25%

PROBLEM 2.48

A cast-in-place concrete silo that weighs 12,500 kip has the following seismic site parameters:

- $S_S = 0.75$
- $S_1 = 0.55$
- $I = 1.0$
- $T = 0.04\ sec$
- Site Class C

Based on the above information, what is the minimum base shear that should be considered for this silo?

(A) 2,250 kip

(B) 2,743 kip

(C) 2,944 kip

(D) 3,135 kip

PROBLEM 2.49

Which of the following nonstructural components should be assigned an importance factor of $I_p = 1.0$?

(A) An egress stairway for a hospital.

(B) An egress stairway for a two-story retail building located in SDC B.

(C) A sprinkler system in a commercial complex with an occupant load of 1,000.

(D) A parapet on a national defense building.

PROBLEM 2.50

The sketch below illustrates a skirt supported vertical welded steel vessel laterally supported on all sides by steel special concentrically braced frames.

The vessel has a height of 90 ft, while the supporting braced frames has a height of 100 ft.

The vessel's effective weight is 90 kip, and the braced frames is 40 kip.

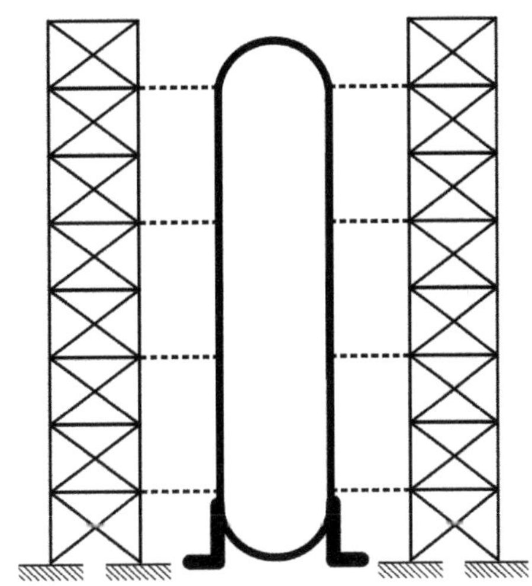

The fundamental period of the vessel is 0.55 sec, and the braced frames is 0.7 sec.

Based on this information, what is the best method to seismically design the supporting structure (i.e., the braced frames)?

(A) The braced frames to be designed using ASCE 7 Chapter 12 or 15.5 using $R = 6$.

(B) The braced frames to be designed using ASCE 7 Chapter 12 or 15.5 using $R = 2$.

(C) The braced frame and the vessel to be modelled together using $R = 2$.

(D) The braced frames and the vessel to be modelled together using $R = 6$.

PROBLEM 2.51

A storage rack consists of two bays with two storage levels, made of cold-formed steel ordinary braced frames. Each rack is $10\ ft$ tall, with bay dimensions of $8\ ft \times 8\ ft$. The total effective weight per rack is $950\ lb$.

Assuming the rack is located in a seismic region of SDC B with the following parameters:

- $S_{DS} = 0.6$
- $S_{D1} = 0.5$
- $T = 0.25$
- $I = 1.0$

What is the required diaphragm (shelf) design force at the upper storage level of the rack system?

(A) $129\ kip$

(B) $255\ kip$

(C) $341\ kip$

(D) $391\ kip$

PROBLEM 2.52

A ground-supported, flat-bottom steel storage tank has an effective seismic weight of $500\ kip$, including the tank shell, roof, and contents.

Based on the above, the maximum calculated base shear for this tank shall not exceed:

(A) $129.3\ kip$

(B) $251.7\ kip$

(C) $288.7\ kip$

(D) $312.5\ kip$

PROBLEM 2.53

Which of the following gusset plate/brace connection(s) provide the least desirable out-of-plan performance?

I. Option I

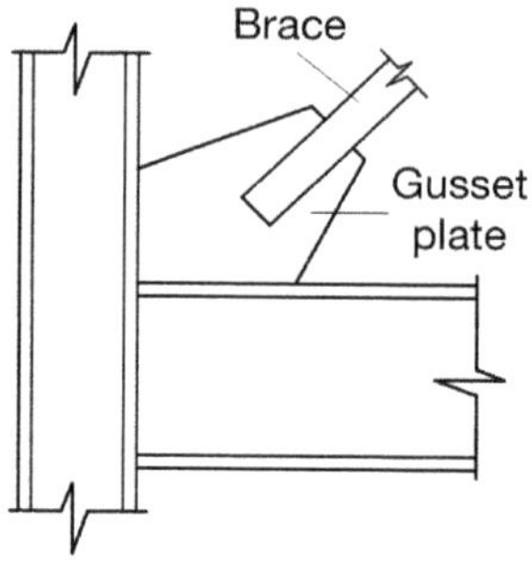

II. Option II

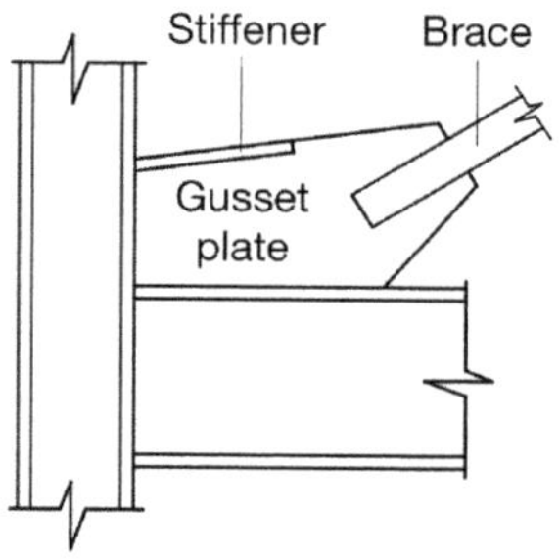

III. Option III

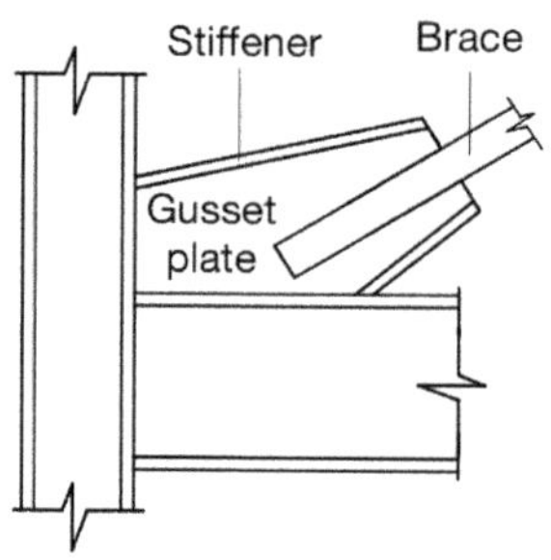

IV. Option IV

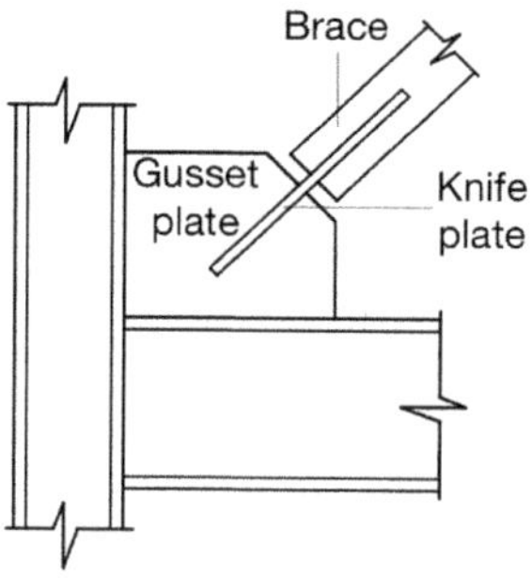

(A) I, II

(B) II, III

(C) III, IV

(D) I, IV

PROBLEM 2.54

Which of the following statements about the panel zone in a steel moment-resisting frame is correct?

(A) The panel zone is designed to resist axial loads from the above floor.

(B) Doubler plates are added to the column flanges to increase flexural capacity of the column.

(C) Doubler plates are added to the column's web if the shear demand exceeds the column web's shear capacity.

(D) The panel zone is not affected by beam moments.

PROBLEM 2.55

According to the 2022 California Building Code, what is the minimum number of project-specific pre-production tests required to determine the capacity of helical piles in a Seismic Design Category D location?

(A) One

(B) Two

(C) Three

(D) Four

BONUS QUESTION

What is the maximum permitted cantilever length allowed for wood diaphragms?

(A) $15\ ft$

(B) $25\ ft$

(C) $35\ ft$

(D) $45\ ft$

Answer Sheet

No.	Our Answer	Your Notes
2.1	A	
2.2	A	
2.3	D	
2.4	C	
2.5	D	
2.6	B	
2.7	C	
2.8	A	
2.9	B	
2.10	D	
2.11	C	
2.12	A	
2.13	D	
2.14	D	
2.15	C	
2.16	B	
2.17	A	
2.18	B	
2.19	D	
2.20	C	
2.21	B	
2.22	B	
2.23	A	
2.24	B	
2.25	A	
2.26	A	
2.27	A	
2.28	B	
2.29	D	
2.30	A	
2.31	B	
2.32	C	
2.33	C	
2.34	D	
2.35	C	
2.36	B	
2.37	A	
2.38	C	
2.39	D	
2.40	A	
2.41	C	
2.42	D	
2.43	B	
2.44	B	
2.45	D	
2.46	D	
2.47	D	
2.48	A	
2.49	D	
2.50	C	
2.51	B	
2.52	C	
2.53	C	
2.54	C	
2.55	B	

SOLUTION 2.1

Using Table 1-1 from *The Essential Guide* for the first $100\,ft$, with $\bar{v}_s = 1{,}250\,ft/s$, this site is classified as **Site Class C** – Very dense soil or soft rock.

> The above information can be found in the ASCE 7 Standard, Section 20.4 Definitions of Site Class Parameters, and Table 20.3-1 Site Classification.

Correct Answer is (A)

SOLUTION 2.2

The correct answer is the **moderately steep slope near a coastal cliff, composed of loose, silty sand with a high-water table.**

The high-water table and loose, silty sand make this slope highly susceptible to failure during an earthquake. The low cohesion in the sand further weakens the soil, making it difficult to form a stable structure or resist sliding forces. All this can lead to slope failure. The lack of cohesion also makes it easier for a sliding face to form, making the slope failure more likely.

For information on slope stability, check *The Essential Guide* Section 1.4.4.

Correct Answer is (A)

SOLUTION 2.3

Referring to page 6 of *The Essential Guide* for the first $> 120\,ft$, with $\bar{s_u} < 1{,}000\,psf$, this site should be classified as **Site Class F.**

> The above information can be found in the ASCE 7 Standard, Section 20.3.1 point 4.

Correct Answer is (D)

SOLUTION 2.4

By observing that $S_1 \geq 0.75$ with the use of Table 1-9 of *The Essential Guide*, you will only need to determine the risk category for this structure. The risk category for a detention facility per Section 6.2.2 page 209 of *The Essential Guide* is category III.

Based on the above information, and taking Table 1-9 mentioned above into account, this facility is assigned an **SDC E.**

> Information provided in this solution can be found in the following sections of codes:
>
> - Information on $S_1 \geq 0.75$ can be found in ASCE 7 Section 11.6 second paragraph.
> - Risk categories can be found in CBC 2022 Section 1604.5.

Correct Answer is (C)

SOLUTION 2.5

We need to find the SDC for this building referring to the methods used in the previous question, then use SDC to compare the height limitations for each SFRS in ASCE 7 Table 12.2-1.

Given a Site Class E for an OSHPD and consulting with Tables 1-2 and 1-3 of *The Essential Guide* to determine values of F_a and F_v:

$$S_{DS} = \frac{2}{3} \times F_a \times S_s = \frac{2}{3} \times 1.2 \times 1.0 = 0.80$$

$$S_{D1} = \frac{2}{3} \times F_v \times S_1 = \frac{2}{3} \times 3.3 \times 0.2 = 0.44$$

The risk category for an acute care facility with no emergency surgery is III. This is based on Section 6.2.2 page 209 of *The Essential Guide.*

Using Table 1-7 & 1-8 of *The Essential Guide* with the above information generates an SDC D.

Based on the above information, with a building height of $100\,ft$ and SDC D, also, taking into account the limitations on some SFRSs as indicated by the CBC 2022 on some systems as greyed out in the appendix of Chapter 2 of *The Essential Guide,* the best system from the above selection in this case is the **Steel and concrete composite partially restrained moment frames.**

Information provided in this solution can be found in the following sections of codes:

- F_a & F_v: CBC 2022 Section 1613.
- SDC Tables: ASCE Tables 11.6-1 & 2.
- Risk categories can be found in CBC 2022 Section 1604.5.
- Check SFRS limitations in CBC 2022 Section 1617A.1.4.C for OSHPD.

Correct Answer is (D)

SOLUTION 2.6

According to Table 12.2-1 of ASCE 7, the most appropriate seismic force resisting system for this scenario is the **intermediate precast shear walls**. While the table initially shows a height limitation of $40\,ft$, note (i) – located below the table – clarifies that this limit may be increased to $45\,ft$ for single-story warehouse buildings (*).

Although Options A and D may also be technically permissible, they are not the most suitable choices for this building type and would likely result in a more complex and less economical solution.

Correct Answer is (B)

(*) Be sure to consult a full-size version of the table to read all applicable notes and exceptions clearly.

SOLUTION 2.7

Use Table 3-1 of *The Essential Guide* and Table 12.8-1 of ASCE 7 to determine the approximate period. The height of the building shall be determined from the lowest level above basement. This information is included in the Notes Section in *The Essential Guide.*

$$T_a = C_t h_n^x$$
$$= 0.02 \times (8 \times 14)^{0.75}$$
$$= 0.69\ sec$$

The above information on fundamental period is in ASCE 7 Section 12.8.2.1.

Information on the location of base is found in ASCE 7 Section C11.2.

Correct Answer is (C)

SOLUTION 2.8

Refer to Section 3.2.4 of *The Essential Guide* to use the fundamental period equation.

$$T = 2\pi\sqrt{\frac{W}{kg}}$$

When all coefficients of this equation remain unchanged, we can compare the new fundamental period with the current one using substitution, as follows:

$$T_{current} = 2\pi\sqrt{\frac{W}{kg}}$$

$$T_{new} = 2\pi\sqrt{\frac{W}{(½)kg}} = \sqrt{2} \times T_{current}$$

An increase in fundamental period results in a decrease in the expected seismic lateral force. This information can be concluded from the shape of the response spectrum where the spectral acceleration S_a decreases with the increase in T.

Information on the calculation of the fundamental period can be found in ASCE 7, Section 13.3.3. The same equation provided for components in that section can also be applied to single-degree-of-freedom systems.

Correct Answer is (A)

SOLUTION 2.9

To improve the ductility of a moment-resisting frame, several strategies can be implemented. First, the frame must comply with strong-column/weak-beam, typically a ratio of 1.2 is used. Additionally, the use of prequalified or tested ductile moment connections ensures reliable inelastic behavior. For example, using Reduced Beam Sections (RBS), to promote plastic hinges in controlled (protected) regions of the beam, is a common method to enhance ductility and shift yielding away from connections.

Applying the above strategies can upgrade a steel frame from an ordinary moment frame to an intermediate or a special one.

Details on RBS and other ductile connection strategies are covered in detail in Chapter 6 of *The Essential Guide.*

On the other hand, Option A – which proposes combining the moment frame with a more ductile system – is not an effective approach. When two lateral force-resisting systems are combined, the overall system performance is governed by the lower-performing component. This concept is discussed in *The Essential Guide*, Section 3.4.1, and in ASCE 7, Section 12.2.3.

Correct Answer is (B)

SOLUTION 2.10

The solution of this question resembles the solution of Problem 1.10 of Exam 1 here with a key difference; when designing supports for members that span two buildings – a connecting bridge in this case – the requirements are more stringent, as outlined in ASCE 7 Section 12.12.4 or Section 2.4.3 of *The Essential Guide.*

Use ASCE 7 Table 12.2-1, along with the risk categories presented in *The Essential Guide* page 209 to collect the following information:

Coefficient	Building 1	Building 2
C_d	2.5	4.5
R	3.0	5.0
Risk Category	II	III
I_e	1.0	1.25
$\delta_{e\ @\ level\ 3}$	3.5 in	3.7 in

$$\delta_{M1} = \frac{1.5R}{C_d} \times \left(\frac{C_d\delta_{max}}{I_e}\right)$$

$$= \frac{1.5\times3.0}{2.5} \times \left(\frac{2.5\times3.5\ in}{1.0}\right)$$

$$= 15.75\ in$$

$$\delta_{M2} = \frac{1.5R}{C_d} \times \left(\frac{C_d\delta_{max}}{I_e}\right)$$

$$= \frac{1.5\times5.0}{4.5} \times \left(\frac{4.5\times3.7\ in}{1.25}\right)$$

$$= 22.20\ in$$

$$\delta_{MT} = \delta_{M1} + \delta_{M2}$$

$$= 15.75 + 22.20$$

$$= 37.95\ in\ (*)$$

Information on buildings' deflection and setbacks are in ASCE 7 Section 12.12.3 and 12.12.4 for connecting supports between two buildings.

Correct Answer is (D)

(*) In addition to this method, diaphragm deflection shall be considered as well.

SOLUTION 2.11

Since this is a rigid diaphragm, the first step is to determine whether any torsional irregularities exist. This assessment is necessary to establish whether an amplification factor should be applied to the accidental torsional moment.

$$\frac{\delta_{max}}{\delta_{ave}} = \frac{5.5}{0.5\times(5.5+1.5)} = 1.6 > 1.4$$

In reference to Figure 2-16 of *The Essential Guide*, this figure shows that this diaphragm has a Type 1.b torsional irregularity. Consequently, and as specified in Table 2-1 of *The Essential Guide*, an amplification factor must be applied to the accidental torsional moment for SDC C. This factor is calculated as follows:

$$A_x = \left(\frac{\delta_{max}}{1.2\ \delta_{ave}}\right)^2 = \left(\frac{5.5}{1.2 \times 3.5}\right)^2 = 1.71 < 3$$

Based on this analysis, the total torsional moment to be applied to the diaphragm is as follows – be sure to include and apply A_x to the accidental torsional moment which is applicable to 5% of the length perpendicular to the load in accordance with page 131 of *The Essential Guide*:

$$M = M_T + A_x \times M_{ta}$$

$$= 75\ kip \times ½\ ft + 1.7 \times 75\ kip \times (5\% \times 18\ ft)$$

$$= 152.25\ kip$$

Information on the above can be found in the following ASCE 7 Sections and Tables:

- Irregularities Table 12.3-1.
- Accidental torsional moment requirements Section 12.8.4.2.
- Torsional moment amplification factor Section 12.8.4.3.

Correct Answer is (C)

SOLUTION 2.12

Table 2-2 of *The Essential Guide* states that Type 1.b (extreme soft story irregularity) is prohibited for OSHPD structures per the CBC 2022 code regulations.

This information can be found in CBC 2022 Section 1617A.1.10.

Correct Answer is (A)

SOLUTION 2.13

ASCE 7 Section 12.10.1 includes a coefficient k (typically ranging from 1 to 2) in the equation for F_{px}, which is used to amplify the diaphragm force to account for higher-mode effects. These higher modes can result in localized force spikes that the first mode alone does not capture. The inclusion of k helps ensure diaphragms are designed for realistic and potentially amplified inertial demands.

Correct Answer is (D)

SOLUTION 2.14

According to Table 7-2 and the notes below it in *The Essential Guide*, the most common seismic failure in older dwellings – particularly light wood-frame homes – is the house coming off its foundation. This typically occurs due to the lack of positive anchorage between the superstructure and the foundation, especially in pre-1970s construction where anchor bolts or sill plate connections were minimal or nonexistent.

This information can also be found in ATC 20 page 36.

Correct Answer is (D)

SOLUTION 2.15

Reference here is made to ASCE 7 Section C12.11.2.2.7. According to this section, when pilasters are incorporated into a wall system, the anchorage force must be calculated considering two-way bending behavior of the wall panels. When the stiffening effect of the pilasters is accounted for, the resulting reaction at the top of each pilaster is governed by the panel's two-way action and is transferred directly to the anchor. In this scenario, the anchor is required to resist the entire tributary load, as illustrated below. It is important to note that this approach represents a conservative assumption, as recommended in ASCE 7.

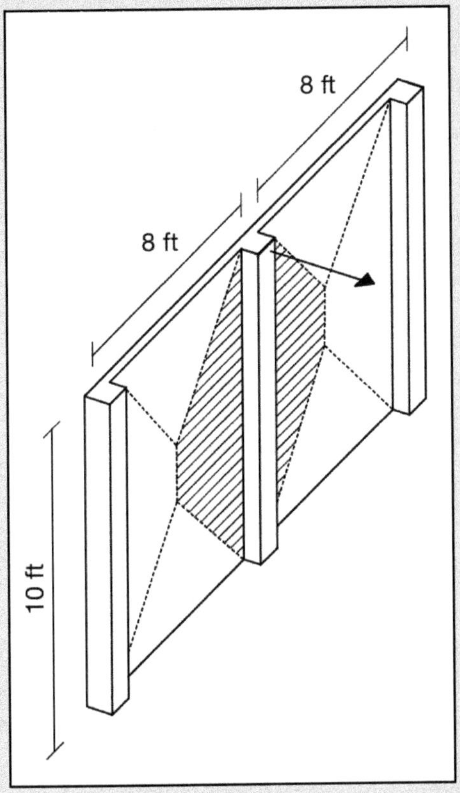

The front elevation view is as follows:

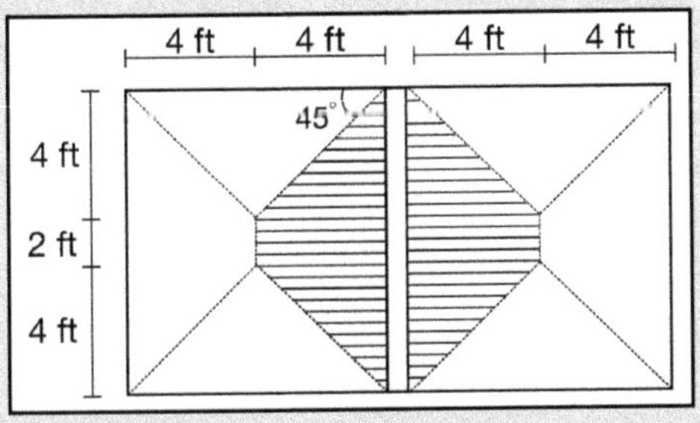

$$Tributary\ Area = 2 \times \left(\frac{10+2}{2} \times 4\right) = 48\ ft^2$$

Correct Answer is (C)

SOLUTION 2.16

Reference is made to *The Essential Guide*, Table 1-2 to determine the required site coefficients, page 209 to determine the risk category, which is III for a power generating station (not backing up category IV facilities), and with the use of Table 1-5 in this case $I_e = 1.25$, and finally Section 3.3.5 to determine the out-of-plane wall forces.

With a Site Class D and $S_s = 1.1 \rightarrow F_a \approx 1.06$:

$$S_{DS} = \frac{2}{3} \times 1.1 \times 1.06 = 0.78$$

$$W_{wall} = \left(150 \frac{lb}{ft^3}\right) \times \left(\frac{9}{12} ft\right) \times 16\ ft \times 22\ ft$$
$$= 39{,}600\ lb$$

$$F_p = 0.4 S_{DS} I_e W_{wall}$$
$$= 0.4 \times 0.78 \times 1.25 \times 39{,}600\ lb$$
$$= 15{,}444\ lb$$

Check against the minimum force as follows:

$$F_{p,min} = 10\% \times W_{wall}$$
$$= 10\% \times 39{,}600$$
$$= 3{,}960\ lb < 15{,}444\ lb \rightarrow OK$$

The information mentioned above can be found in the following codes:

- F_a: ASCE 7 Table 11.4-1.
- Risk categories can be found in CBC 2022 Section 1604.5.
- Out-of-plane forces: ASCE 7 Section 12.11.1.

Correct Answer is (B)

SOLUTION 2.17

Reference is made to *The Essential Guide*, Section 3.2.2 for the effective weight calculation, and to Section 3.3.2 for base shear calculation.

The effective weight for this building in the specified direction should include the slab and half the height of shear walls that are not resisting in that direction. See below:

$$W_{slab} = 150\frac{lb}{ft^3} \times \frac{12}{12} ft \times 25\,ft \times 15\,ft$$
$$= 56{,}250\,lb$$

$$W_{walls} = 2 \times \left(150\frac{lb}{ft^3} \times \frac{9}{12} ft \times \frac{14\,ft}{2} \times 10\,ft\right)$$
$$= 15{,}750\,lb$$

$$W_{total} = 56{,}250 + 15{,}750 = 72{,}000\,lb$$

$$V = C_s V = 0.85 \times 72{,}000 = 61{,}200\,lb$$

The information mentioned above can be found in the following codes:

- Base shear: ASCE 7 Section 12.8.1.
- Effective seismic weight: ASCE 7 Section 12.7.2.

Correct Answer is (A)

SOLUTION 2.18

These conditions are outlined in ASCE 7, Section 12.3.4.2 and Table 12.3-3. In this case, Option B applies to a two-story building, which states that the ground level must resist more than 35% of the total base shear (refer to point a of Section 12.3.4.2). This section then directs you to Table 12.3-3 under the same category, which specifies that the loss of a column base connection must not result in more than a 33% reduction in the lateral load-resisting capacity of the story.

Correct Answer is (B)

SOLUTION 2.19

When precast diaphragms are used, multiple connection methods can be considered. In this case, **Option D** is the most appropriate cross-section. It shows an anchor plate embedded in the grout-filled core, secured by a rebar welded to the plate. This setup effectively connects the slab to the wall via a weld to the embedded angle.

On the other hand, **Option A** illustrates a precast slab doweled directly into the grout core. While this option may work, but it lacks critical detailing required for proper load transfer.

The rest of the options are weak, and by proper visual inspection, shear is not going to be appropriately transferred to the core of the wall.

It is advisable to explore reliable online resources or consult the reference list provided in *The Essential Guide* for more anchorage cross-sections as this is an important and common topic.

Correct Answer is (D)

SOLUTION 2.20

Reference is made to *The Essential Guide*, Section 3.4.2, which states that when the lower SFRS has a lower response modification factor, each system may use its own for analysis. In this case, the lower system is a cantilevered column system, as defined in ASCE 7 Table 12.2-1, with a response modification factor of $R = 1.5$. The upper system consists of light wood frame walls sheathed with wood structural panels rated for shear resistance, with a response modification factor of $R = 7$. Therefore, the upper two stories may be analyzed using their own response modification factor of $R = 7$.

The information mentioned above can be found in the following codes:

- SFRS horizontal combination: ASCE 7 Section 12.2.3.1.
- ASCE 7 Table 12.2-1 for the various SFRS system coefficients.

Correct Answer is (C)

SOLUTION 2.21

Check the aspect ratio for walls:

$$\frac{h}{l} = \frac{9}{4} = 2.25 < 3$$

$$\frac{h}{l} = \frac{9}{7} = 1.29 < 3$$

The above indicates that the walls will experience a combined flexural and shear deflection as explained in Section 3.2.3 of *The Essential Guide.*

The Essential Guide Table 4-1 explains that we shall use the rigidities provided in Table 3-6 for this purpose for cantilever walls – see the below grey text box if you do not own a copy of *The Essential Guide* for a long solution.

There are two wall types, 4 ft long and 7 ft long both are 9 ft tall. Using Table 3-6 (*):

$$R_{4ft} = 0.019$$

$$R_{7ft} = 0.08$$

$$\bar{x} = \frac{0 + 11 \times 0.08 + 22 \times (0.019 + 0.08)}{0.08 + 0.08 + 0.08 + 0.019} = 11.8$$

$$\bar{y} = \frac{0 + 10 \times (0.019 + 0.019)}{0.019 + 0.019 + 0.019} = 6.7$$

The following equations outline how to calculate the above relative rigidities:

$$R_{4ft} = \frac{1}{4\left(\frac{h}{l}\right)^3 + 3\left(\frac{h}{l}\right)} = \frac{1}{4\left(\frac{9}{4}\right)^3 + 3\left(\frac{9}{4}\right)} = 0.019$$

$$R_{7ft} = \frac{1}{4\left(\frac{h}{l}\right)^3 + 3\left(\frac{h}{l}\right)} = \frac{1}{4\left(\frac{9}{7}\right)^3 + 3\left(\frac{9}{7}\right)} = 0.08$$

Correct Answer is (B)

(*) You may notice that in some resources, these values are presented as 10 times larger to emphasize rigidity. This is often referred to as relative rigidity. Both approaches are valid as long as the same method is used consistently throughout the solution. However, the values provided here are more accurate.

SOLUTION 2.22

Check the aspect ratio for walls:

$$\frac{h}{l} = \frac{9}{36} = 0.25 < 0.3$$

$$\frac{h}{l} = \frac{9}{40} = 0.225 < 0.3$$

$$\frac{h}{l} = \frac{9}{60} = 0.15 < 0.3$$

$$\frac{h}{l} = \frac{9}{72} = 0.125 < 0.3$$

Based on this, Table 4-1 of *The Essential Guide* states that wall length should be used as an indicator of relative rigidity, especially that all walls have the same thickness. If the wall thicknesses vary, the wall top view cross-sectional area should be used instead.

$$\bar{x} = \frac{0 \times 60 + 72 \times 40}{40 + 60} = 28.8$$

$$\bar{y} = \frac{0 \times 72 + 60 \times 36}{36 + 72} = 20$$

Correct Answer is (B)

SOLUTION 2.23

Typically, when the effective seismic weights are the same for all floors, the diaphragm design force will be equivalent to the vertically distributed base shear at the roof level → 98 kip.

This concept is illustrated in Example 3.7 of *The Essential Guide.* You can also verify this yourself by assuming a uniform floor weight of W here and applying the equation provided in Section 3.3.4 of *The Essential Guide.*

The information mentioned above can be found in ASCE 7 Section 12.10.1.1.

Correct Answer is (A)

SOLUTION 2.24

The calculated fundamental period appears to be slightly higher than what is typical for a building of this height. Therefore, we will verify the maximum allowable fundamental period by comparing it to the approximate period T_a multiplied by the coefficient C_u, as outlined in Tables 3-1 and 3-2 of *The Essential Guide.*

According to Table 3-1, there are two methods to estimate the approximate fundamental period T_a:

$$T_a = C_t h_n^x = 0.016 \times 100^{0.9} = 1.0\, sec$$

Or

$$T_a = 0.1N = 0.1 \times 10 = 1.0\, sec$$

Based on this, and using Table 3-2, $C_u = 1.4$, and hence $T_{max} = C_u T_a = 1.4 \times 1.0 = 1.4\, sec.$

Using Section 3.3.3 of *The Essential Guide*:

$$C_{vx} = \frac{w_x h_x^k}{\sum_{i=1}^{n} w_i h_i^k}$$

Where:

$$k = 0.75 + 0.5T = 0.75 + 0.5 \times 1.4 = 1.45$$

$$C_{v4} = \frac{w h_2^{1.45}}{\sum_{i=1}^{n} w h_i^{1.45}} = \frac{20^{1.45}}{\sum_{i=1}^{n} h_i^{1.45}}$$

Don't forget that $h_{x\, or\, i}$ in this equation represents the height from base to the required floor number.

$$= \frac{20^{1.45}}{\begin{bmatrix} 10^{1.45}+20^{1.45}+30^{1.45}+40^{1.45}+50^{1.45} \\ + \\ 60^{1.45}+70^{1.45}+80^{1.45}+90^{1.45}+100^{1.45} \end{bmatrix}}$$

$$= \frac{77.0}{3{,}648.1}$$

$$= 0.021$$

The information mentioned above can be found in ASCE 7 Sections:

- T_a and C_u: Section 12.8.2, Equations 12.8-7 & 8 Table 12.8-1 & 2.
- C_{vx}: Section 12.8.3, Equation 12.8-12.

Correct Answer is (B)

SOLUTION 2.25

The information mentioned above can be found in ASCE 7 Section 12.14.7.6 as 10% of the weight of the wall only.

Correct Answer is (A)

SOLUTION 2.26

Refer to Section 3.2.4 of *The Essential Guide* to use the fundamental period equation.

$$T = 2\pi\sqrt{\frac{W}{kg}}$$

The stiffness k can be calculated using Table 3-5 of *The Essential Guide* for three types/orientations of columns:

$$I_{12\times24} = \frac{12 \times 24^3}{12} = 13{,}824\, in^4$$

$$I_{24\times12} = \frac{24 \times 12^3}{12} = 3{,}456\, in^4$$

$$I_{14\times14} = \frac{14 \times 14^3}{12} = 3{,}201.3\, in^4$$

$$k_{12\times24} = \frac{12EI_{12\times24}}{h^3}$$

$$= \frac{12\times4.2\times10^3 \frac{kip}{in^2}\times13{,}824\, in^4\times\frac{ft^2}{144\, in^2}}{(12\, ft)^3}$$

$$= 2{,}800\, kip/ft$$

$$k_{24\times12} = \frac{12EI_{24\times12}}{h^3}$$

$$= \frac{12\times4.2\times10^3 \frac{kip}{in^2}\times3{,}456\, in^4\times\frac{ft^2}{144\, in^2}}{(12\, ft)^3}$$

$$= 700\, kip/ft$$

$$k_{14\times14} = \frac{12EI_{14\times14}}{h^3}$$

$$= \frac{12\times4.2\times10^3 \frac{kip}{in^2}\times3{,}201.3\, in^4\times\frac{ft^2}{144\, in^2}}{(12\, ft)^3}$$

$$= 648.4\, kip/ft$$

$$\sum k = 6k_{12\times24} + 2k_{24\times12} + k_{14\times14}$$

$$= 6 \times 2{,}800 + 2 \times 700 + 648.4$$

$$= 18{,}848.4\, kip/ft$$

$$W = 22\,ft \times 18\,ft \times 1\,ft \times 150\,pcf$$
$$= 59{,}400\,lb\ (59.4\,kip)$$

$$T = 2\pi\sqrt{\frac{59.4\,kip}{18{,}848.4\,kip/ft \times 32.174\,ft/sec^2}}$$
$$= 0.062\,sec$$

Information on the calculation of the fundamental period can be found in ASCE 7, Section 13.3.3. The same equation provided for components in that section can also be applied to single-degree-of-freedom systems.

Correct Answer is (A)

SOLUTION 2.27

In this configuration, the sheathing panels in **Option A** do not share a common framing element, making it the weakest of all the options. Without a continuous load path, shear transfer is compromised.

In contrast, **Options B & C** both have a single sheathing panel that overlaps and ties the diaphragm to the shear wall, creating a continuous and effective load path for shear forces.

In **Option D**, while a continuous sheathing panel is absent, a shear angle is provided to transfer loads. Although this connection may be less effective for resisting wind uplift – due to the reliance on nail withdrawal – it can still perform adequately for shear resistance, which is the focus of this question.

Correct Answer is (A)

SOLUTION 2.28

For this question, refer to *The Essential Guide*, Section 4.5.2, which presents three load combinations that include seismic loads (load combination 8, 9 and 10). These can be applied here, and we will evaluate them to determine which one aligns with the options provided. Also, no overstrength is required, and hence E_h will be used for the horizontal seismic effects.

Additionally, as noted in *The Essential Guide*, page 148 – and also in Table 2-1 of Chapter 2 – redundancy factor (ρ) must be included in the load combination for this element and should equal to 1.3 due to the extreme torsional irregularity mentioned in the question (see ASCE 7 Section 12.3.4.2).

Load combination 8 per *The Essential Guide* page 148 (ASCE 7 Section 2.4.5):

$$1.0D + 0.7E_v + 0.7E_h$$

Where:

$$E_v = 0.3S_{av}D = 0.3 \times 0.3D = 0.09D$$

$$E_h = \rho Q_E = 1.3Q_E$$

$$L = L\,\&\,S = 0$$

The final equation is therefore:

$$1.0D + 0.7 \times 0.09D + 0.7 \times 1.3Q_E$$

$$\rightarrow 1.063D + 0.91Q_E$$

Load combination 9:

$$1.0D + 0.525E_v + 0.525E_h + 0.75L + 0.75S$$

$$1.0D + 0.525 \times 0.09D + 0.525 \times 1.3Q_E + 0.75L + 0$$

The final equation is therefore:

$$1.05D + 0.68Q_E + 0.75L$$

Load combination 10:

$$0.6D - 0.7E_v + 0.7E_h$$

$$0.6D - 0.7 \times 0.09D + 0.7 \times 1.3Q_E$$

The final equation is therefore:

$$0.54D + 0.91Q_E$$

Based on this, Option B is similar to the final load combination 9.

The information mentioned above can be found in ASCE 7 Sections 2.4.5, 12.4 and 12.3.4.2.

Correct Answer is (B)

SOLUTION 2.29

ASCE 7 Section C12.4.2.2 permits the vertical seismic load effect to be taken as *zero* for buildings assigned to SDC B.

Correct Answer is (D)

SOLUTION 2.30

Equations for flexible diaphragms can be taken from Figure 4-2 of *The Essential Guide* as follows:

$$M_{chord\,@\,10ft} = \frac{Px}{2} = \frac{10\ kip \times 10\ ft}{2} = 50\ kip.ft$$

$$T_{chord\,@\,10ft} = \frac{M_{chord\,@\,10ft}}{D} = \frac{50\ kip.ft}{36\ ft} = 1.4\ kip$$

Correct Answer is (A)

SOLUTION 2.31

The diaphragm chord in this case consists of horizontal reinforcements in the wall near or at the diaphragm level as shown in **Option B**. Using the wood top plate as the chord **(Option D)** is inappropriate due to the greater stiffness of the masonry wall.

Proper chord development in this case requires the diaphragm to be adequately anchored to the shear wall.

Option A represents blocking for the main frame. The blocking may be discontinuous and therefore cannot resist tension forces.

Option C serves to connect the main diaphragm framing to the wall top plate and does not function as a chord.

Correct Answer is (B)

SOLUTION 2.32

Refer to ASCE 7 Table 12.6-1 second point, line 4, and Table 12.8-2 along with equation 12.8-7 (this equation is also found in *The Essential Guide* Table 3-1).

In order to use the ELF procedure per ASCE 7 Table 12.6-1 for structures exceeding $160\,ft$ with no structural irregularities:

$$T < 3.5T_s$$

$$T_s = \frac{S_{D1}}{S_{DS}} = \frac{0.3}{0.5} = 0.6$$

This means that, to be eligible to use the ELF procedure: $T < 3.5 \times 0.6 = 2.1$

Use Equation 12.8-7 to determine T for a concrete moment frame, along with Table 12.8-2 for the frame's coefficients.

Based on this, the maximum height can be determined, as follows:

$$T_a = C_t h_n^x$$

$$2.1 = 0.016 \times h_n^{0.9}$$

$$\rightarrow h_n = \left(\frac{2.1}{0.016}\right)^{1/0.9} = 225.7\ ft$$

Correct Answer is (C)

SOLUTION 2.33

Begin by calculating the load share carried by the braces on the right side, assuming the relative rigidity of the left frame is $R_{left} = 2$ and the right frame is $R_{right} = 1$.

$$= \frac{1}{1+2} \times 15 = 5\ kip$$

Analyze the pinned joint at the upper right of the right-side frame:

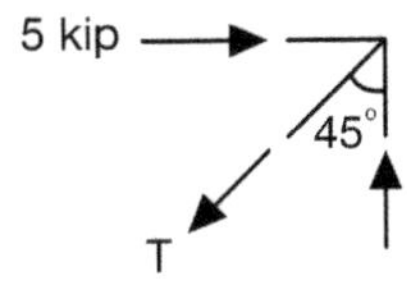

$$T = \frac{5\,kip}{\cos(45)} = 7.1\,kip$$

Correct Answer is (C)

SOLUTION 2.34

We have three collectors in the direction of the provided load. Use the tributary area method to determine the load on each:

Left side collector:

$$V = \frac{500\,lb/ft \times 15\,ft}{2} = 3{,}750\,lb\ (3.75\,kip)$$

$$v_{collector} = \frac{3.75\,kip}{25\,ft} = 0.15\,klf$$

$$V_{max} = 0.15\,klf \times 10\,ft = 1.5\,kip$$

Right side collector:

$$V = \frac{500\,lb/ft \times 20\,ft}{2} = 5{,}000\,lb\ (5.0\,kip)$$

$$v_{collector} = \frac{5.0\,kip}{25\,ft} = 0.2\,klf$$

$$V_{max} = 0.2\,klf \times 10\,ft = 2.0\,kip$$

Middle collector:

$$V = \frac{500\frac{lb}{ft} \times 15\,ft}{2} + \frac{500\frac{lb}{ft} \times 20\,ft}{2}$$

$$= 8{,}750\,lb\ (8.75\,kip)$$

$$v_{collector} = \frac{8.75\,kip}{25\,ft} = 0.35\,klf$$

$$v_{shear\,wall} = \frac{8.75\,kip}{10\,ft} = 0.875\,klf$$

$$\rightarrow V_{max} = 0.35\,klf \times 10\,ft = 3.5\,kip$$

This is also depicted in a collector force diagram as explained in solutions 1.34 and 1.35 as follows (*):

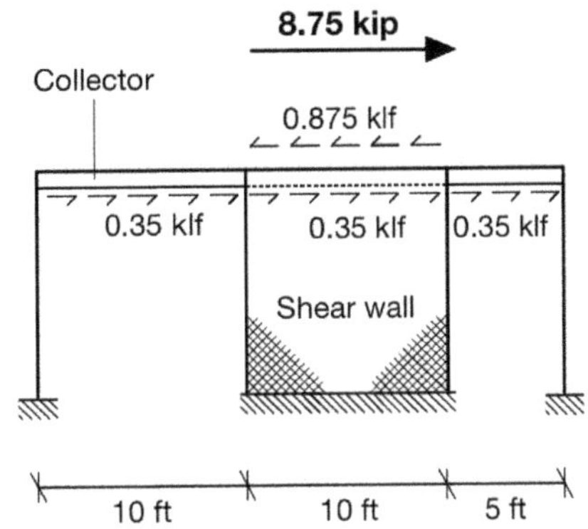

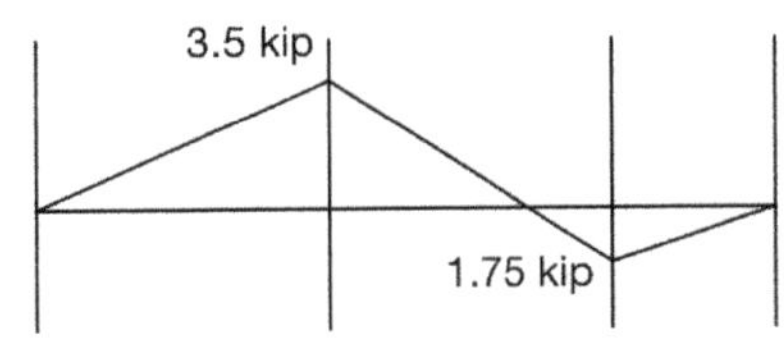

The maximum of the three forces is experienced by the middle collector and is therefore = 3.5 kip

Correct Answer is (D)

(*) When constructing this diagram, it is essential to treat it as a force bookkeeping exercise. The diagram should start at zero from one side and closes at zero on the other side.

From the collector diagram, note that the collector on the left side is in compression as it exerts force on the shear wall (depicted as positive on the diagram), while the collector on the right is in tension as it pulls the middle shear wall in the shown direction.

More examples and theory explanation are available in *The Essential Guide*.

SOLUTION 2.35

Since this is a flexible diaphragm, the lateral force of 50 kip can be disturbed over the shear wall lines of resistance using the tributary area method as follows:

$$Area\ to\ the\ left = 15 \times 15$$
$$= 225\ ft^2\ (31\%)$$

$$Area\ to\ the\ right = 20 \times 25$$
$$= 500\ ft^2\ (69\%)$$

The below shows the calculation for forces on the right-hand-side collector and the middle collector – the middle collector calculation is a bit tricky, but not difficult, since it has two different-sized diaphragms on each of its sides.

Right-hand side collector:

$$V = \frac{50\ kip \times 69\%}{2} = 17.25\ kip$$

$$v_{collector} = \frac{17.25\ kip}{25\ ft} = 0.69\ klf$$

$$V_{max} = 0.69\ klf \times 10\ ft = 6.9\ kip$$

Middle collector:
This is better explained with the use of the following sketch, the calculation on the sketch follows:

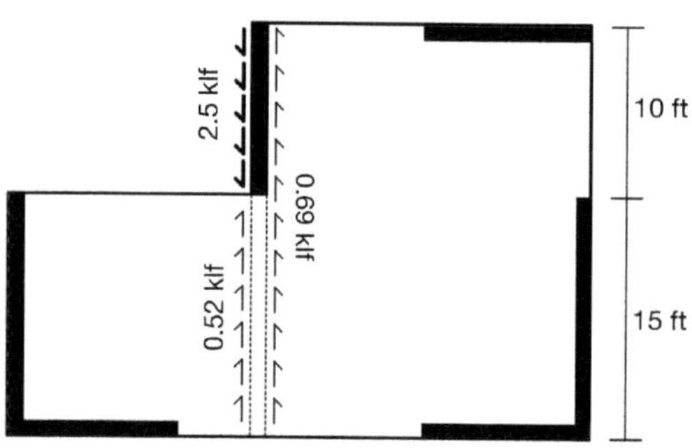

The unit shear on the wall is calculated as follows:

$$V = \frac{50\ kip \times 31\%}{2} + \frac{50\ kip \times 69\%}{2} = 25\ kip$$

$$v_{shear\ wall} = \frac{25\ kip}{10\ ft} = 2.5\ klf$$

The unit shear on each side of the collector is calculated as follows:

$$V_{left\ side} = \frac{50\ kip \times 31\%}{2} = 7.75\ kip$$

$$v_{left\ side} = \frac{7.75\ kip}{15\ ft} = 0.52\ klf$$

$$V_{right\ side} = \frac{50\ kip \times 69\%}{2} = 17.25\ kip$$

$$v_{left\ side} = \frac{17.25\ kip}{25\ ft} = 0.69\ klf$$

Place all the above results on an the elevation plan for the shear wall and the collector:

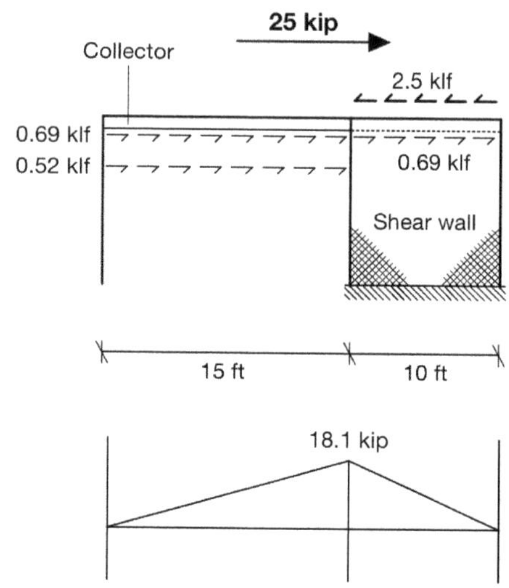

$$V_{max} = (0.69 + 0.52) \times 15 = 18.1\ kip$$

Or

$$= (2.5 - 0.69) \times 10 = 18.1\ kip$$

The maximum value occurs at the middle collector and is therefore 18.1 kip.

Correct Answer is (C)

SOLUTION 2.36

Since this is a flexible diaphragm, the lateral force of 50 kip can be disturbed over the shear wall lines of resistance using tributary area as follows:

$$Area\ A\ to\ B = 12 \times 12 = 144\ ft^2\ (28.6\%)$$

$$Area\ B\ to\ C = 18 \times 20 = 360\ ft^2\ (71.4\%)$$

Based on this, each diaphragm share of this load can be determined and applied to a simple beam between B and C, and a cantilever beam between A and B (assuming B is fixed in this case). See below:

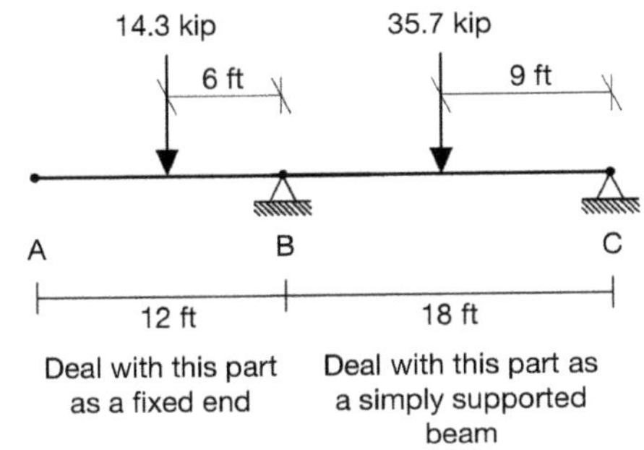

$$R_{wall\ B} = 14.3 + \frac{35.7}{2} = 32.15\ kip$$

Correct Answer is (B)

SOLUTION 2.37

Given that the diaphragm is flexible and non-uniform between supports, it can be modeled as a simply supported beam spanning lines A and B, as explained in Figure 4-1 of *The Essential Guide.*

First, determine the center of the resultant load of 50 kip which would be located at the centroid of the diaphragm.

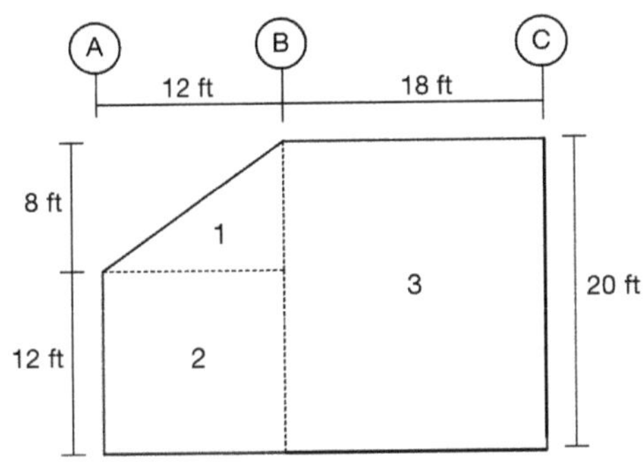

Take axis at line A as datum:

$$\bar{X} = \frac{8\times(½\times8\times12)+6\times(12\times12)+21\times(18\times20)}{(½\times8\times12)+(12\times12)+(18\times20)}$$

$$= 16\ ft$$

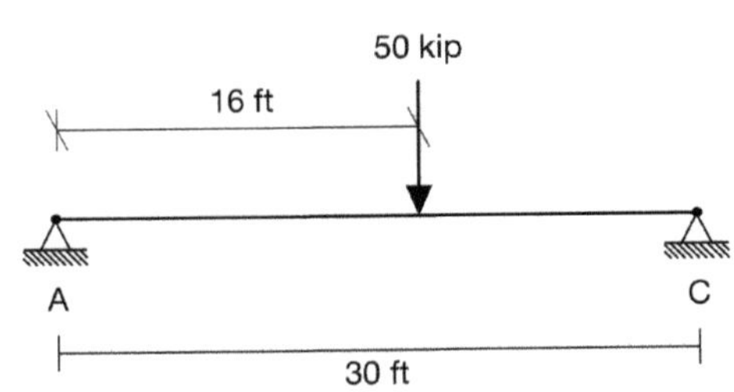

Take moment around support A to find reaction at support C:

$$R_{wall\ C} = \frac{50\ kip \times 16}{30} = 26.7\ kip$$

Correct Answer is (A)

SOLUTION 2.38

If you own a copy of *The Essential Guide*, check Section 4.3.3 for detailed explanation of the methodology used here along with more examples.

ASCE 7 Section 12.8.4.2 requires that accidental torsion (M_{ta}) due to displacement of CM perpendicular to the direction of loading to be considered.

$$e_{CM\ displacment} = 5\% \times 30\ ft = \mp 1.5\ ft$$

$$M_{ta} = \mp 1.5 \times 100 = \mp 150\ kip.ft$$

The inherent torsional moment (M_T) is calculated as follows:

$$M_T = 5 \times 100 = 500\ kip.ft\ \circlearrowright$$

Using the North direction as the Y-axis, apply Equation 1a from Section 4.3.3 of *The Essential Guide*. Note that distances (x) and (y) in the equation represent the distance from walls (R_y) and (R_x) to CR along the (x) and (y) directions, respectively.

$$F_{y,A} = \frac{R_{y,A}}{\sum R_y} \times F_y \mp \frac{(M_T \mp M_{ta}) \times (x) \times R_{y,A}}{\sum R_y(x)^2 + \sum R_x(y)^2}$$

Before applying the above equation, it is important to sketch a simple diagram to understand the direction of loading. This will help in determining how to apply the negative and positive signs in the equation. See diagram that follows in the next page.

Based on this diagram, it can be observed that the direction of the torsional moment is clockwise. This indicates that the direction of loading that is contributed from M_T on wall A will align with the direction of lateral loading, suggesting that the first sign will be positive.

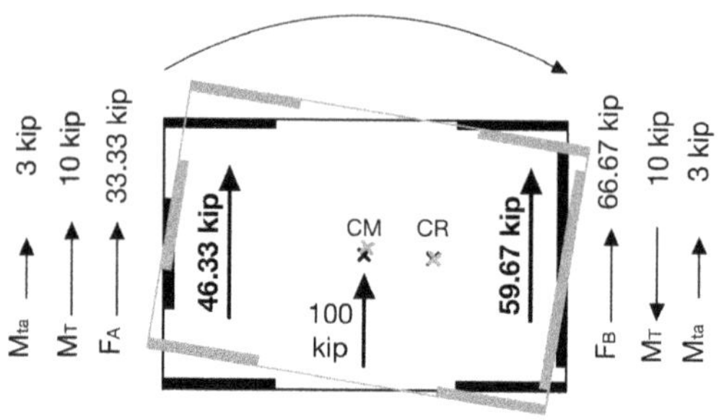

$$F_{y,A} = \frac{R_{y,A}}{\sum R_y} \times F_y + \frac{(M_T \mp M_{ta}) \times (x) \times R_{y,A}}{\sum R_y(x)^2 + \sum R_x(y)^2}$$

To determine the second sign for the $(M_T \mp M_{ta})$ portion of the equation, you may use two methods:

- If the first sign is positive, the second sign will also be positive. If the first sign is negative, so will the second. This maximizes load sharing on the wall for accidental moment.
- Alternatively, move the CM away from the CR towards the desired wall, which will result in a positive sign in this case.

$$F_{y,A} = \frac{R_{y,A}}{\sum R_y} \times F_y + \frac{(M_T + M_{ta}) \times (x) \times R_{y,A}}{\sum R_y(x)^2 + \sum R_x(y)^2}$$

$$= \frac{1}{1+2} \times 100 + \frac{(500 + 150) \times 20 \times 1}{(1 \times 20^2 + 2 \times 10^2) + 4 \times (1 \times 10^2)}$$

$$= 33.33 + \frac{13{,}000}{1{,}000}$$

$$= 46.33\ kip\ (*)$$

Correct Answer is (C)

(*) The rest of the forces in the SN and WE directions for the rest of the walls are solved and sketched below for your benefit:

Wall B:

$$F_{y,B} = \frac{R_{y,B}}{\sum R_y} \times F_y + \frac{(M_T \mp M_{ta}) \times (x) \times R_{y,B}}{\sum R_y(x)^2 + \sum R_x(y)^2}$$

$$= \frac{2}{1+2} \times 100 - \frac{(500 - 150) \times 10 \times 2}{(1 \times 20^2 + 2 \times 10^2) + 4 \times (1 \times 10^2)}$$

$$= 66.67 - \frac{7{,}000}{1{,}000}$$

$$= 59.67\ kip$$

Any wall in the horizonal direction:

Because all walls in this direction share the same rigidity and are all equidistant from the CR, we can use one equation for all which is Equation 2a of Section 4.3.3 from *The Essential Guide*.

$$F_{x,i} = \mp \frac{(M_T + M_{ta})(y)R_{x,i}}{\sum R_y(x)^2 + \sum R_x(y)^2}$$

$$= \mp \frac{(500 + 150) \times 10 \times 1}{1{,}000}$$

$$= \mp 6.5\ kip$$

The below sketch summarizes all forces that are transferred to all walls:

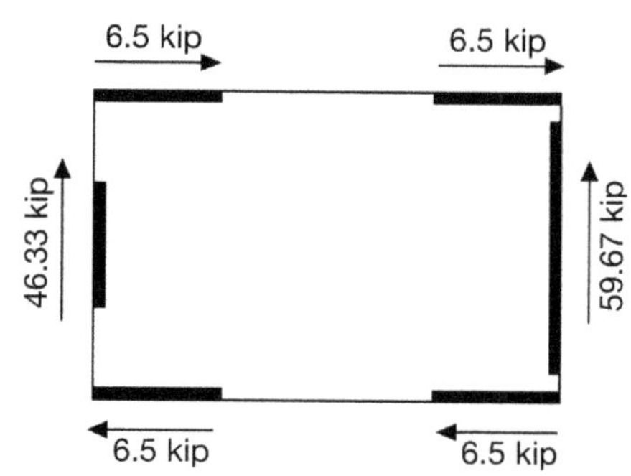

SOLUTION 2.39

In a well-designed EBF, the system is intentionally configured so that inelastic behavior (ductile energy dissipation) occurs in the short segment of the horizontal beam known as the "link beam". This segment is designed to yield in flexure and shear under lateral loads, providing controlled ductile behavior and preventing failure in other components.

This is explained in more detail in *The Essential Guide* Section 6.6.2.

Correct Answer is (D)

SOLUTION 2.40

Reference is made in this solution to CBC 2022, Table 2306.3(1). It is important to note that this table provides the allowable capacities for stapled wood shear walls, unlike SDPWS which provides nominal value for nailed shear walls.

The necessary calculation involves multiplying the values provided in the table by the length indicated in the sketch associated with this question as follows:

Wall	v_a (plf)	Length (ft)	V_a (kip)
1	155	12	1.86
2	235	9	2.12
3	375	12	4.5
			8.48

Correct Answer is (A)

(*) For additional information, consult *The Essential Guide* Section 6.3.4 where this method is described with key examples.

It is also necessary to consider the notes below this table, such as using a different wood grade than Douglas Fir-Larch, or varying wood moisture content, and how these variables can influence the final shear capacity.

SOLUTION 2.41

Inspectors cannot make a unilateral decision to shut down an entire hospital. You may restrict certain operations, limit new admissions, or close specific areas. The decision to post a hospital as "Unsafe" must be made with extreme caution, as evacuating large hospitals with thousands of patients – many of whom may be on life support – is complex and risky. In such situations, consulting with the appropriate staff or emergency team is the best course of action before making any of the above decisions. This information can be found in ATC-20 page 108.

Correct Answer is (C)

SOLUTION 2.42

Statement D is not necessarily true. Nonparticipating walls are defined by their lack of involvement in the seismic (lateral) resisting system, but they can be either reinforced or unreinforced depending on their purpose. Reinforcement may still be provided for vertical load bearing or out-of-plane resistance.

You can refer to Section 6.4.2 of *The Essential Guide* for further information on the definitions, characteristics, and design considerations of participating and nonparticipating walls.

For information on nonparticipating walls see TMS 402 Chapter 7.

Correct Answer is (D)

SOLUTION 2.43

You can refer to Section 7.2.3 of *The Essential Guide*, which states that essential facilities should be prioritized for detailed assessment. This information is also found in Chapter 13 of the ATC, particularly in the first two paragraphs, where examples are provided.

Based on these references, the parking garage of the dispatch center is among the least likely to receive a detailed inspection.

For this question, if you do not have a copy of *The Essential Guide*, you may refer to ATC 20 Chapter 13.

Correct Answer is (B)

SOLUTION 2.44

The method for determining the appropriate load case for the nailed diaphragm is explained in *The Essential Guide*, page 222.

This diaphragm has a staggered panel arrangement, where the continuous panel edges are parallel to the framing, and the applied load is also parallel to the framing direction. According to CBC 2022 Table 2306.2(1), this configuration corresponds to Load Case 4.

While this diaphragm includes the use of staples with $1\,½\,in$ 16 gauges, based on the values in CBC 2022 Table 2206.2(1), and the details provided in the question, the allowable unit shear capacity is $\mathbf{200\,plf}$.

You can find the information in this question in CBC 2022 Table 2306.2(1).

Correct Answer is (B)

SOLUTION 2.45

The equation from Section 5.3.3 of *The Essential Guide* (Equation 13.3-1 from ASCE 7), along with Table 13.5-1 of ASCE 7, can be used to calculate the lateral seismic design force.

The following parameters are collected for Cantilever parapets from ASCE 7 Table 13.6-1.

$$a_p = 2.5$$

$$R_p = 2.5$$

$$W_p = 9.5 \times 4 = 38\,plf$$

$$I_p = 1.5$$

The second floor means the diaphragm/ceiling of the first level, while the hospital has four equal height levels, hence:

$$\frac{z}{h} = \frac{1}{4} = 0.25$$

Apply Equation 13.3-1 of ASCE 7:

$$F_p = \frac{0.4a_pS_{DS}W_p}{\left(\frac{R_p}{I_p}\right)}\left(1 + 2\frac{z}{h}\right)$$

$$= \frac{0.4 \times 2.5 \times 0.6 \times 38}{\left(\frac{2.5}{1.5}\right)}\left(1 + 2 \times \frac{1}{4}\right)$$

$$= 20.5\,plf$$

Check minimum and maximum values for the design force:

$$0.3S_{DS}I_pW_p \le F_p \le 1.6S_{DS}I_pW_p$$

$$0.3S_{DS}I_pW_p = 0.3 \times 0.6 \times 1.5 \times 38 = 10.3\,plf < 20.5\,plf \rightarrow ok$$

$$1.6S_{DS}I_pW_p = 1.6 \times 0.6 \times 1.5 \times 38 = 54.7\,plf > 20.5\,plf \rightarrow ok$$

Hence an $F_p = 20.5\,plf$ governs.

You can find the information presented in this solution in the following sections/tables of ASCE 7:

- Equations 13.3-1, 2 & 3 for seismic design forces.
- Table 13.5-1 for nonstructural components coefficients.

Correct Answer is (D)

SOLUTION 2.46

The equation from Section 5.3.3 of *The Essential Guide* (Equation 13.3-1 from ASCE 7), along with Table 13.6-1 of ASCE 7, can be used to calculate the lateral seismic design force. In this solution however, you need to pay close attention to the comments below Table 13.6-1.

The following parameters are collected for air handlers from ASCE 7 Table 13.6-1.

$$a_p = 2.5$$

$$R_p = 6$$

Apply Equation 13.3-1 of ASCE 7 – or the equation in *The Essential Guide* for the use on roofs of structures:

$$F_p = 3 \times \frac{0.4a_pS_{DS}W_p}{\left(\frac{R_p}{I_p}\right)}$$

$$= 3 \times \frac{0.4 \times 2.5 \times 0.3 \times 7}{\left(\frac{6}{1.0}\right)}$$

$$= 1.05\ kip$$

Check minimum and maximum values for the design force:

$$0.3S_{DS}I_pW_p \le F_p \le 1.6S_{DS}I_pW_p$$

$$0.3S_{DS}I_pW_p = 0.3 \times 0.3 \times 1.0 \times 7 = 0.63\ kip < 1.05\ kip \rightarrow ok$$

$$1.6S_{DS}I_pW_p = 1.6 \times 0.3 \times 1.0 \times 7 = 3.36\ kip > 1.05\ kip \rightarrow ok$$

However, note *b* of ASCE 7 Table 13.6-1 states that the design force shall be taken as $2F_p$ if the nominal clearance (air gap) between the equipment support frame, and the restraint is greater than $0.25\ in$.

→ Use $2F_p = 2.1\ kip$

You can find the information presented in this solution in the following sections/tables of ASCE 7:

- Equations 13.3-1, 2 & 3 for seismic design forces.
- Table 13.6-1 for nonstructural components coefficients.

Correct Answer is (D)

SOLUTION 2.47

Reference in this question is made to Table 5-2 of *The Essential Guide* where it states that 25% of weight of all equipment NOT fastened to access floors shall be included in the operating weight of those floors.

You can find the information presented in this solution in ASCE 7 Section 13.5.7.1 for access floors.

Correct Answer is (D)

SOLUTION 2.48

Reference in this question is made to following sections and tables from *The Essential Guide*:

- Figure 5-6 which presents one of our custom diagrams that takes you through the steps of solving nonstructural components and nonbuilding structure questions with confidence and ease.
- Tables 1-2 of *The Essential Guide* to determine values of F_a.

Using the diagram presented in Figure 5-6, and since this silo is rigid given that $T < 0.06\ sec$:

$$V = 0.3S_{DS}WI_e$$

Given a Site Class C and consulting with Table 1-2 of *The Essential Guide* to determine the value of F_a:

$$S_{DS} = \frac{2}{3} \times F_a \times S_s = \frac{2}{3} \times 1.2 \times 0.75 = 0.60$$

$$V = 0.3 \times 0.6 \times 12{,}500 \times 1 = 2{,}250\ kip$$

Information provided in this solution can be found in the following sections:

- ASCE 7 Table 11.4-1 for F_a
- ASCE 7 15.4.2 for the calculation of seismic forces for rigid nonbuilding structures with $T < 0.06$.

Correct Answer is (A)

SOLUTION 2.49

This question refers to Section 5.3.1 of *The Essential Guide*. According to this section, and consistent with ASCE 7 Section 13.1.3, the Importance Factor should be assigned as 1.5 when the component is essential for **life safety** or for the **continued operation of essential facilities** following an earthquake (such as Risk Category IV buildings). In this context:

Option A: Egress systems are critical for life safety and post-earthquake functionality → $I_p = 1.5$.

Option B: While the building is not Risk Category IV, egress stairways are still considered essential for life safety in all occupancies → $I_p = 1.5$.

Option C: Fire suppression systems are specifically listed as life safety systems in ASCE 7 → $I_p = 1.5$.

Option D: While a national defense building is Risk Category IV (see page 210 of *The Essential Guide*), the parapet is not essential to the continued operation of the facility, nor is it a component directly related to life safety (unless its failure could obstruct egress or cause injury in a way that impairs emergency operations) → $I_p = 1.0$.

You can find the information presented in this solution in ASCE 7 Section 13.1.3.

Correct Answer is (D)

SOLUTION 2.50

This question uses *The Essential Guide,* Section 5.4.2 page 180, or Figure 5-6, which presents a simplified design process for nonbuilding structures. For further detail, see ASCE 7, Section 15.3.

According to the referenced design diagram, the vessel qualifies as a nonbuilding structure with a weight greater than 25% of the total.

With the steel vessel having a fundamental period $T > 0.06\ sec$, permits the supporting structure to be modelled along with the vessel in a combined modeled with the use of the lesser R of both, in which case $R = 2$.

You can find the information presented in this solution in ASCE 7 Section 15.3.

Correct Answer is (C)

SOLUTION 2.51

Per *The Essential Guide,* Section 5.4.3, and as directed by ASCE 7 Section 15.4.1 point 4a, the vertical distribution of lateral seismic forces for nonbuilding structures is similar to normal buildings.

Refer to ASCE 7 Table 15.4-1 to determine R for the specified SFRS. For a cold-formed steel storage rack ordinary braced frame → $R = 3$.

Using Table 5-4 & 5 of *The Essential Guide*:

$$T_s = \frac{S_{D1}}{S_{DS}} = \frac{0.5}{0.6} = 0.83 > T\ (= 0.25)$$

$$\rightarrow C_s = \frac{S_{DS}}{\left(\frac{R}{I_e}\right)} = \frac{0.6}{\left(\frac{3}{1}\right)} = 0.2$$

$$C_{s,min} = 0.044 S_{DS} I_e = 0.044 \times 0.6 \times 1 \approx 0.03$$

$$\rightarrow \text{Use } C_s = 0.2$$

$$V = C_s W = 0.2 \times (2 \times 950) = 380\ lb$$

Use the equation presented in Section 3.3.3 for vertical distribution as follows:

$$F_x = C_{vx} V$$

$$C_{vx} = \frac{w_x h_x^k}{\sum_{i=1}^{n} w_i h_i^k} \quad \text{where } k = 1 \text{ for } T < 0.5$$

	h_x	h_x^k	w_x	$w_x h_x^k$	C_{vx}	F_x
2	20	20	950	19,000	0.67	255
1	10	10	950	9,500	0.33	125
				28,500		**380**

The information presented above can be found in ASCE 7 Sections:

- 15.4.1 point 4a for vertical distribution of nonbuilding structures along with Section 12.8.3
- R: Table 15.4-1.
- $C_{s,min}$: Equation 15.4-1.
- C_{vx}: Section 12.8.3.

Correct Answer is (B)

SOLUTION 2.52

This question references Table 5-9 of Chapter 5 from *The Essential Guide*, which specifies the permissible maximum base shear for such tanks in this scenario:

$$V = Wtan(30) = 500\tan(30) = 288.7\ kip$$

You can find the information presented in this solution in ASCE 7 Section 15.7.6.1.5(a).

Correct Answer is (C)

SOLUTION 2.53

When a bracing member buckles out-of-plane, plastic hinges form in the gusset plate. For these hinges to dissipate energy effectively and avoid premature failure, they must be free to rotate. If this rotational capacity is restricted, the gusset plate may fracture after only a few seismic cycles.

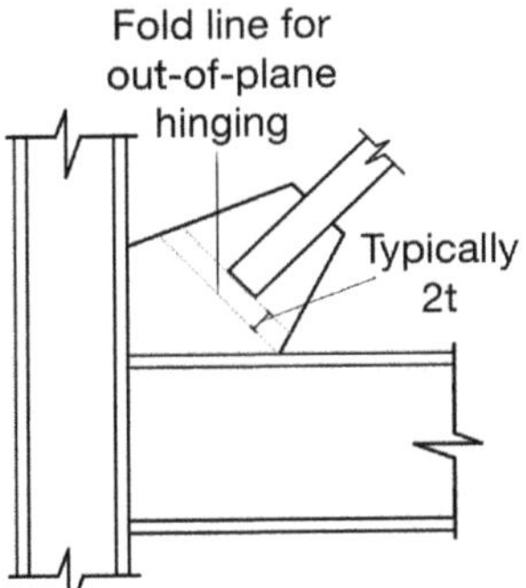

To ensure sufficient rotational freedom, the brace should terminate at a distance away from the beam-column intersection, allowing the gusset plate to form a fold line. A typical recommendation is twice the gusset plate thickness. See Option I in the preceding sketch – this configuration is preferred because it facilitates proper out-of-plane deformation.

In this context, stiffener plates are generally discouraged. Over-stiffening the gusset plate can inhibit its ability to yield and rotate, increasing the likelihood of fracture. Gusset plates should not be overly strong or stiff; instead, they should be detailed to yield in a controlled, ductile manner.

Option II is also considered desirable. It includes a stiffener, but it terminates before the folding line, preserving the gusset plate's ability to rotate. In most cases, such a stiffener is optional, and its inclusion should be carefully evaluated.

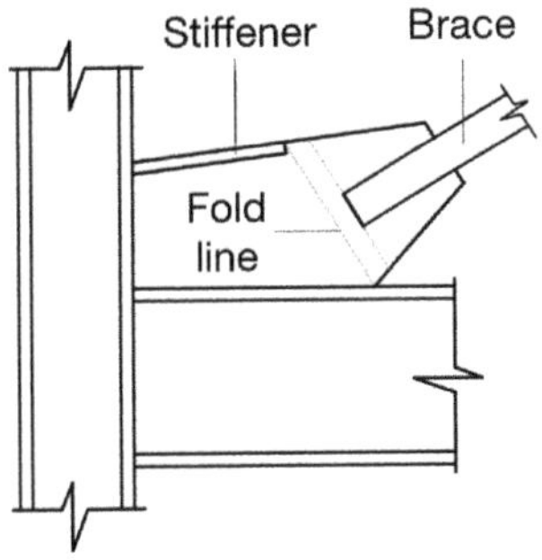

Option III, however, introduces excessive stiffening across the folding zone. The brace also extends across it. This prevents the required out-of-plane rotation, making the gusset more prone to brittle fracture. As such, **Option III is not recommended for seismic applications**.

Option IV represents a knife plate connection, which, while potentially acceptable for in-plane behavior, limits out-of-plane flexibility. **This makes Option IV least desirable** for out-of-plane behavior.

Correct Answer is (C)

SOLUTION 2.54

In a steel moment-resisting frame, when a beam meets a column, the area of the column web between the top and bottom beam flanges is called the panel zone. During seismic loading, the beam transfers moment to the column, which in turn causes shear forces to develop in this panel zone. If the shear demand in the panel zone exceeds what the column web can handle, doubler plates – steel plates welded to one or both sides of the column web are added to increase shear strength and stiffness at the joint.

Correct Answer is (C)

SOLUTION 2.55

The CBC 2022 requires a minimum of two project-specific pre-production tests for each soil profile to determine the capacity of helical piles in SDC D, E and F. This along with other requirements are summarized in Table 6-1 of *The Essential Guide.*

You can find the information presented in this solution in CBC 2022 Section 1810.3.1.5.1.

Correct Answer is (B)

BONUS QUESTION SOLUTION

The SDWPS 2021 states that the maximum cantilever length for wood diaphragms is $35\,ft$. This information is provided in the Notes section of *The Essential Guide.*

You can find the information presented in this solution in SDPWS 2021 Section 4.2.6.

Correct Answer is (C)

NOTES

Notes:

Your Feedback Matters – Make Sure You Share it With Others

Good day,

As you reach the final pages of this book, I would like to express my sincere gratitude for choosing it as your guide to aid you in your journey toward success in the PE exam. I have poured countless hours into meticulously crafting the material in these pages.

Your opinion matters greatly in helping others discover the value of this resource. If you found this book beneficial, kindly consider leaving your feedback on the platform that you bought it from - like Amazon. Your words will not only acknowledge the hard work invested into producing this book but will also guide future readers in their quest for quality study materials.

Remember, your review is more than just feedback; it is a beacon for those seeking reliable resources. Your support can make a significant difference, ensuring that this book continues to assist aspiring professionals on their path to success.

Thank you for being a part of this journey, and I personally appreciate your commitment to sharing your experience with others.

Jacob Petro

PETRO PUBLICATIONS

www.ingramcontent.com/pod-product-compliance
Ingram Content Group UK Ltd.
Pitfield, Milton Keynes, MK11 3LW, UK
UKHW051452290926
15471UKWH00008B/1167